DATA ANALYSIS
CASE STUDIES IN BUSINESS STATISTICS
VOLUME II

PRACTICAL DATA ANALYSIS

CASE STUDIES IN BUSINESS STATISTICS
VOLUME II

SECOND EDITION

Peter G. Bryant
Marlene A. Smith
College of Business and Administration
and
Graduate School of Business Administration
University of Colorado at Denver

Irwin McGraw-Hill

Boston   Burr Ridge, IL   Dubuque, IA   Madison, WI   New York   San Francisco   St. Louis
Bangkok   Bogotá   Caracas   Lisbon   London   Madrid
Mexico City   Milan   New Delhi   Seoul   Singapore   Sydney   Taipei   Toronto

Irwin/McGraw-Hill

A Division of The McGraw·Hill Companies

PRACTICAL DATA ANALYSIS: CASE STUDIES IN BUSINESS STATISTICS VOLUME II
Copyright ©1999 by The McGraw-Hill Companies, Inc. All rights reserved. Previous edition ©1995 by Richard D. Irwin, Inc. Printed in the United States of America. Except as permitted under the United States Copyright Act of 1976, no part of this publication may be reproduced or distributed in any form or by any means, or stored in a data base or retrieval system, without the prior written permission of the publisher.

This book is printed on acid-free paper.

1 2 3 4 5 6 7 8 9 0 DOC/DOC 9 3 2 1 0 9 8

ISBN 0-256-23872-3

Vice president/Editor-in-chief: *Michael W. Junior*
Publisher: *Jeffrey J. Shelstad*
Senior sponsoring editor: *Scott Isenberg*
Developmental editor: *Wanda J. Zeman*
Marketing manager: *Zina Craft*
Senior project manager: *Susan Trentacosti*
Senior production supervisor: *Heather D. Burbridge*
Cover Designer: *Steven Vena/SrV Unlimited Design*
Compositor: *Carlisle Communications, Ltd.*
Typeface: *10/12 Times Roman*
Printer: *R. R. Donnelley & Sons Company*

Library of Congress Cataloging-in-Publication Data

Bryant, Peter G.
   Practical data analysis : case studies in business statistics /
Peter G. Bryant and Marlene A. Smith.—2nd ed.
     p. cm.
   Includes bibliographical references (p.).
   ISBN 0-256-23871-5 (vol. 1). —ISBN 0-256-23872-3 (vol. 2). —
ISBN 0-07-365488-4 (vol. 3)
   1. Commercial statistics—Case studies. 2. Commercial statistics-
–Data processing. 3. Statistics—Data processing. I. Smith,
Marlene A. II. Title.
HF1017.B79 1999
650'.01'5195—dc21   98-43159

http://www.mhhe.com

To L.S.B. and L.G.B.

In memory of Reuben and Charlotte Aschenbach

And he who is versed in the science of numbers can tell of the regions of weight and measure, but he cannot conduct you thither.

Kahlil Gibran
The Prophet

Was man nicht nützt, ist eine schwere Last.
Johann Wolfgang Goethe
Faust I, 684

Some of the blocks to dealing comfortably with numbers . . . are due to quite natural psychological responses to uncertainty, to coincidence, or to how a problem is framed. Others can be attributed to . . . romantic misconceptions about the nature and importance of mathematics.

John Allen Paulos
Innumeracy

INTRODUCTION FOR THE STUDENT

INTRODUCTION

This book contains a collection of *cases* or situations. Most cases contain some *data*. In others, we ask you to gather data or figure out how you *would* gather the data. Presumably the data can shed some light on the situation. Your job is to decide what light the data shed. To do this, you will have to

- Understand the situation and context.
- Choose an appropriate technique to summarize or analyze the data.
- Decide what you think the data have to say.

Then you will have to communicate your findings, in a way that's effective in the context of the situation described, whether that be a presentation to a board of directors, a memo to your boss, or something else.

CASES

In some fields of education—law or medicine, for example—you often "solve" a case by applying your knowledge to the situation given in a systematic way, and if you apply your knowledge correctly, you arrive at a single correct, well-defined answer, like a legal opinion or a diagnosis. Different people applying the course material correctly will arrive at the same answer.

In *business* education, cases usually play a different role: they are vehicles to raise issues, point out inherent conflicts, and stimulate debate. At the end of the debate, you may have settled some issues, but others remain open. Your answers or conclusions will not necessarily agree with those of your colleagues, for they will have brought different assumptions, interpretations, and solution methods to the debate.

We expect that you will *debate* these cases. We designed them that way. You may debate various aspects of the situation:

- The analysis method to use.
- The quality of the data.
- The proper interpretation of the analysis results.
- The recommendations or conclusions to draw.

For example, you may find you agree with your colleagues about the appropriate analysis of the data, but disagree about whether the data are of a quality that justifies a conclusion. Sometimes the analyses will be easy, sometimes hard. Sometimes a single method of attack seems obvious; at other times the situation yields to various different approaches. Some cases seem to require relatively sophisticated methods for full solution, but simple techniques will lead to at least part of the message. All the cases can profit from discussion and debate, though; they are not purely statistics or computational homework exercises.

REMARKS

Your instructor will specify the formats he or she wants you to use for your presentation and written summary, and will guide any debate. We often ask students to prepare a one-page summary (perhaps augmented by a chart or graph) and to come to class prepared to present and/or defend their findings in a general class discussion. This works particularly well when you prepare your report in a group. You will find it takes a lot of work to get a group of three to five people to agree on what the message in the data is! When you have to boil it down to one page, it puts even more pressure on you to know exactly why you did what you did and what you think it means. Often, other students will use different techniques and/or come to different conclusions, and you will have to defend your approach and conclusions to them. That's part of the fun and the frustration: statistics is not meant to be a sterile exercise at the end of a textbook chapter. It's supposed to be techniques for real people to better inform themselves about real situations by using real data. That sometimes means ambiguity, controversy, assumptions, disagreement and hard work. We think you'll find it helpful, worthwhile, and relevant, though. That's the message from our students who've tried it this way.

There's no one "right" approach to any of these cases, although often we had a particular approach in mind when we wrote it. Almost every time we assign one of them, though, some student points out a valid alternative we hadn't thought about. And just because the case is in the book doesn't mean we think the results will be overwhelming. Sometimes, the data don't really have much to say. That happens in real life, and it may happen here, too.

The cases are given in random order, so you can't tell what technique we had in mind by where the case is in the book or by how you approached the case next to it.

Most of the cases include computer files with data, and most of the time you will want to use a computer but not always. The presence of a computer file doesn't necessarily mean we think that you should use a computer. For those instances in which you would *like* to use a computer, we have prepared data files in a variety of formats. Your instructor will give you more information about acquiring and using the data files.

In our classes, we place a lot of emphasis on how well students write up or present their findings. In practice, a good analysis may influence nobody if it isn't presented well, and we think you should make it a habit, right from the start, to put as much energy into your presentation as you do into the analysis. We don't particularly advocate fancy, color graphics or other "fashionable" approaches. Good hand-drawn graphs can be wonderfully effective, too. The key is to communicate clearly. We discuss report writing more in the next section.

We hope you have fun with the cases. If you come up with an example from your own life or career that you think others would enjoy or profit from, please let us know. Maybe the next edition will have room for it!

Peter G. Bryant
Marlene A. Smith

ACKNOWLEDGMENTS

Many colleagues have reviewed individual cases and the manuscript of *Practical Data Analysis.* We thank specifically:

Sung K. Ahn, *Washington State University*
Randy Anderson, *California State University at Fresno*
Bridget G. Heidemann, *Seattle University*
Raj Jagannathan, *University of Iowa*
Jerzy J. Letkowski, *Western New England College*
Walter J. Mayer, *University of Mississippi*
Dale McFarlane, *Oregon State University*
J. B. Orris, *Butler University*
Paul Paschke, *Oregon State University*
John R. Pickett, *Georgia Southern University*
Kipling M. Pirkle, *Washington and Lee University*
Al Schainblatt, *San Francisco State University*
Milo Schield, *Augsburg College*
William L. Seaver, *University of Tennessee*
Herbert Spirer, *University of Connecticut*
Jeffrey W. Steagall, *University of North Florida*
Stanley A. Taylor, *California State University at Sacramento*
Betty M. Thorne, *Stetson University*
Thomas K. Tiemann, *Elon College*
Wayne Winston, *Indiana University at Bloomington*

We acknowledge also the founders, organizers and participants in the annual conference entitled *Making Statistics More Effective in Schools of Business* for their hard work and interest in changing the way in which we teach statistics in our business classrooms.

For specific permissions we thank the American Statistical Association for permission to reproduce items from *The American Statistician;* the *Rocky Mountain News* for copyright permission; and Stanley D. Elofson for assisting us in obtaining copyright permission for the 1992 Arapahoe County property assessment study. Further acknowledgments are listed with individual cases.

Several friends, colleagues, and relatives helped in various ways. Edward J. Conry advised us on the preparation of our prospectus. Lucinda Bryant, Louise Bryant, and Lynwood Bryant read the book and commented on content, style, and presentation. Kathryn E. Chmelir and Adriano Pimenta helped us with research, data input and analysis, and other related tasks. The staff at Irwin/McGraw-Hill, including Scott Isenberg and Wanda Zeman, have also been helpful.

Finally, we thank our students and other contributors to the cases in the book. Where appropriate, we acknowledge them individually in the source notes of their respective cases. All of our students contributed indirectly to this project: they told us what did and didn't work, complained about bad cases (and rightly so!), and had fun with the good ones.

Peter G. Bryant
Marlene A. Smith

CONTENTS

	PREPARING WRITTEN CASES AND BUSINESS REPORTS	II–xv
	THE DATA DISK	II–xxi
CASE 26	Dave's Truck Washing Service	II–1
CASE 27	Housing Prices—I	II–5
CASE 28	Housing Prices—II	II–9
CASE 29	Books	II–13
CASE 30	ELISA Plate Uniformity	II–17
CASE 31	Violence on TV—I	II–25
CASE 32	Overdue Bills	II–27
CASE 33	The Brazing Operation	II–29
CASE 34	Leading Indicators	II–33
CASE 35	Interrupting Tina	II–37
CASE 36	Violence on TV—II	II–43
CASE 37	Commercial Property	II–47
CASE 38	AIDS Rates	II–51
CASE 39	Software Service Calls	II–53
CASE 40	Titanic Shipping Company	II–55

CASE 41	Campaign Statistics	II–59
CASE 42	Hospital Charges	II–61
CASE 43	Coal Testing	II–67
CASE 44	Performance Evaluations	II–71
CASE 45	Amchem's Tax Credit	II–83
CASE 46	Natural Gas Prices	II–87
CASE 47	Westar Beverage Sales	II–89
CASE 48	Price-Fixing Cases	II–91
CASE 49	Property Crimes	II–93
CASE 50	Rockville Physicians' Group	II–97

PREPARING WRITTEN CASES AND BUSINESS REPORTS

The cases in this book require you to communicate your statistical results somehow. You may discuss your results in class with your instructor and your other classmates. You may give formal classroom presentations. Often, you will prepare a written report. Writing and speaking effectively are important to your professional career. Here are some thoughts on writing.

There are many ways to organize a business memo, but formal structure is probably less important than clear, concise communication. Writing the summary of your results will probably be just as difficult as doing the computations. Our students say it takes about as long to write the results as it does to finish the statistical work. You might want to plan accordingly.

WRITING STYLE

Once you are done with the computer work, you must interpret these findings for someone, perhaps a manager or your boss.

Avoid Jargon

Write so that *anyone* can understand it. Don't use technical language to intimidate or overwhelm the recipient. Unless you are writing to someone well-versed in statistical techniques, don't use statistical and mathematical jargon in your report. If you were the vice president of marketing, and had never studied statistics, which of the following two summaries would you find most valuable?

- The regression model is given by: SALES = 13,000 + 5*ADVERTISING, which means that the *y*-intercept is $13,000 and the slope is $5.

• According to the model: (*a*) if the firm does no advertising, sales will be $13,000, and (*b*) each additional dollar spent on advertising will bring in five additional dollars in sales.

Your instructor and the textbook will use jargon. You will see specific statistical terms like "least squares" and "p-values" in class, but don't presume that the *reader* of your memo understands them. You must translate statistical concepts, methods, and outcomes for the uninitiated. By virtue of spending time in your statistics class, you may forget that certain statistical concepts don't exist in most people's vocabularies. Here are some guidelines.

• Don't use words or phrases in your report unless you used them before signing up for statistics.
• Would your next door neighbor understand the essence of your report?
• What would the editor of your local newspaper think about publishing your report in tomorrow's newspaper?

Use the Active Voice

Use the active voice. "I ate my dinner" says more than "The dinner was eaten by me," and says it better.

Be Clear

Say what you mean, unambiguously. "Nothing is capable of being well set to music that is not nonsense,"[1] is a quotable epigram, but does it mean "Nothing is capable of being well set to music-that-is-not-nonsense"? Or does it mean "Nothing-that-is-not-nonsense is capable of being well set to music"? If you mean "only nonsense can be set to music," say so.
Similarly,

"My face looks like a wedding cake left out in the rain."[2]

is great poetry, but leaves room for various interpretations.

ORGANIZATION

Introductory Material

To help your reader, provide some background information, such as a statement of the problem or situation and a description of the data available to answer this question. Since your report involves statistical analyses, you might also provide initial descriptive statistics (e.g., means and/or standard deviations) of the more important variables, unless that information would simply distract from your point. Such background information puts the problem in perspective and eases the reader into the upcoming material.

---

[1] Joseph Addison (1711), quoted in *The Columbia Dictionary of Quotations*. Copyright © 1993 by Columbia University Press. All rights reserved.

[2] W. H. Auden (1907–73), quoted in Humphrey Carpenter, *W. H. Auden*, pt. 2, ch. 6 (1981). *The Columbia Dictionary of Quotations*. Copyright © 1993 by Columbia University Press. All rights reserved.

Good Grammar

An effective report uses good grammar, correct spelling and appropriate punctuation. You can't convince a reader that your statistical results are valid if your writing is poor. Sloppy, misspelled, or disorganized reports send a message: you don't think the report is very important. Brush up on your writing skills. It's fun and valuable to write effectively. Word processors may help. A spelling checker is useful, too.

Junked-Up Appendixes

Consider putting supporting statistical documentation, such as graphs, tables, and other statistical output, at the end of your written report. Within the report, refer to these appendixes (e.g., "See Exhibit II") for guidance. On the other hand, if one particular table or graph contains the essence of the point you're making, put it with the text, where your reader can see it quickly. Ask yourself: If I put it in with the text, will it distract the reader more than if the reader has to turn to an appendix? A critical graph belongs in with the text. A table giving relatively minor support to your argument belongs in an appendix.

Don't append every statistical printout you produced. Include the *important* ones. If *you* can't decide which is important and which is not, how can your readers? In fact, if you haven't decided which are important, you're not done with your analysis yet. "Don't append a statistical exhibit if you don't refer to it in the report, and, of course, don't refer to an exhibit that's not there" might be a good guideline.

Length

We generally limit reports to one or two pages. Your instructor may impose different limits, but write as succinctly as you can. Present the essence of your statistical findings, not a comprehensive validation of every step of your work. Managers won't wade through pages and pages of information. It's easy to overwhelm a reader who wants nothing more than a summary of the major findings, but why do so?

You must identify the fine line between too much detail and not enough detail. Don't make your report so short that it's deceptive, but don't bore or intimidate your intended reader with unnecessary details, either. You, the reporter of the data, must decide what to include.

Even if your report is a lengthy project, we suggest that you make the first page an "executive summary," written for someone who has never had a statistics class, and doesn't care to have statistics explained. Include only the *important* results and implications derived from the data, and any necessary caveats or limitations of your findings. For instance, you might not trust your results because of a small sample size, or some confidence interval might suggest inaccuracy of a point estimate. Try to limit this executive summary to one page, single-spaced. It's hard to do!

To start you off, we've included sample reports on the next three pages. Preparing reports will become increasingly straightforward with practice. Good luck!

A PRETTY LOUSY BUSINESS REPORT

TO: Tom Jones, Director of Personnel

FROM: Jana Smith, Statistician

DATE: 5/19

SUBJECT: Analysis of Personnel Data

Per your request, I have analyzed the data on sick days and the company's new wellness program. The results are summarized below.

The regression model suggests that there is a statistically significant relationship between the two variables. The correlation between dollars contributed to the wellness program and absenteeism is a positive 0.63. The standard error is 0.073. The R-squared statistic on the simple regression model is 56%, which is pretty good for cross-sectional data. Moreover, the F-statistic measuring the statistical significance of the model as a whole is 54.90, indicating a good model.

When analyzing the relationship by gender, there is no statistically significant impact here. The p-value on the categorical variable called GENDER (see EXHIBIT II) is .46. This means that absenteeism does not seem to be correlated with the gender of the employee. On the other hand, it just might be a multicollinearity problem with the two independent variables.

I'd be happy to assist in any further analysis of this data at your request.

A PRETTY GOOD BUSINESS REPORT

TO: Tom Jones, Director of Personnel

FROM: Jana Smith, Statistician

DATE: 5/19

SUBJECT: Analysis of Personnel Data

As you asked, I have analyzed the data on sick days and the company's new wellness program. Here are my results.

Data

The personnel department provided a sample of 125 randomly chosen employee files. From those files, we obtained:

- the company's payment for that employee's participation in our wellness program,

- that employee's absentee record (measured in number of absent days) over the past two years, and

- the gender of the employee.

72% of the sample group was female.

Results

- On average, our employees missed 15 days of work in the first year. The typical fluctuation around the average of 15 days was seven days.

- In the second year, after starting the wellness program, the average number of absent days declined to 10 days, and the typical fluctuation also decreased to 2 days. We committed about $50 per employee to the wellness program last year.

- A statistical model (see EXHIBIT below) of the relationship between dollars committed to wellness, absenteeism, and gender indicates that the wellness program has a statistically significant relationship to absenteeism. The model suggests that:

    - each additional dollar committed to our wellness program was associated with a two hour decline in absenteeism, and

    - there is no statistically significant relationship between gender and absenteeism.

The model would generally be considered a statistically strong one, since 56% of the variation in absenteeism is explained by the model. Although this leaves 44% of the variation unexplained, it is difficult to do much better with the type of data available for this analysis.

Recommendation

Because of the decline in absenteeism after the institution of the wellness program, I recommend that the program be continued.

Limitations

Consider redoing this study in another year with a larger sample size. It is not clear whether one year is enough to observe the full benefits of the wellness program. Also, there are some troubling aspects of the statistical results that might be alleviated with a larger sample size. For instance, many of the standard errors in the model (that is, measures of the accuracy of the estimates) are quite large in my opinion.

I'd be happy to answer any further questions that you might have about my analysis or report.

```
                         EXHIBIT

The regression equation is
ABSENT = 12.2 - 0.26 DOLLARS + 0.07 GENDER

Predictor       Coef       Stdev       t-ratio        p
Constant        12.2       3.37        3.62           0.00
DOLLARS         -0.26      0.04        6.50           0.00
GENDER          0.07       0.25        0.28           0.80

s = 7.9    R-sq = 56.2%   R-sq(adj) = 55.5%

Analysis of Variance

SOURCE          DF         SS          MS         F         p
Regression      2          9770        4885       78.3      0.000
Error           122        7614        62.4
Total           124        17384
```

THE DATA DISK

Most of the cases in *Practical Data Analysis* require you to analyze a data set. A printed version (hard copy) of the data set may be found at the end of each case. The printed version lets you get a quick, informal sense of the data. At some point, you will probably want to use a statistical or spreadsheet software package for more formal analyses of these numbers. Some data sets are small enough to enter the data yourself. For others, that task would be very time consuming.

Your instructor has access to the data files for the cases in *Practical Data Analysis*. She or he will distribute those files to you in a form that is appropriate for the computing environment at your school. You may also receive instructions on how to copy the files for your own use and storage, and how to feed the data into one or more software programs.

Once you receive the data in machine-readable form from your instructor, you should be able to get up and running quickly.

PRACTICAL DATA ANALYSIS

CASE STUDIES IN BUSINESS STATISTICS
VOLUME II

CASE 26

DAVE'S TRUCK WASHING SERVICE

Dave is the owner of the Truck Washing Service (TWS). TWS provides cleaning services to truckers and trucking operations within a moderately sized metropolitan area. Dave is interested in understanding the factors that influence monthly fluctuations in total sales of TWS.

TWS offers two types of services. Type 1 sales are from Dave's mobile truck washing service. Dave also provides a mobile chemical treatment service for removing decals and logos from trucks, called type 2 sales. Most of TWS revenues come from the type 1 operations. Occasionally, there will be an upcoming auction of used trucks in town. Trucks are generally cleaned up prior to auctions, and Dave can pick up extra business at auction time.

Dave does all of the marketing for TWS and also handles the administrative and financial side of the business. When necessary, he occasionally works in the field. Dave started out with only one other full-time employee. Later, he was able to add an additional part-time employee to the payroll.

Dave has recently become sensitive to the old adage, "While the cat's away, the mice will play." Sales seem to drop pretty dramatically when Dave is on vacation. Dave has also recently enrolled in an MBA program, which reduces the amount of time that he can commit to his business. During the semester, Dave suspects that business drops off.

The data file is named TWS. This file contains 46 monthly time-series observations and eight variables. The time-series begins in May. The definitions of the variables on that file are SALES, the total monthly sales of TWS in dollars; SALES2, the monthly sales of type 2 in dollars; TEMP, the average monthly temperature (degrees Fahrenheit);

These data are real, but the name of the company has been changed. The data were made available to us by David S. Larson and Carmin Hunter-Siegert. Carmin is with the Colorado Student Obligation Bond Authority. We would also like to thank Dave and Carmin for coming to several of our classes to discuss the data set and their study with our students.

and PREC, the total monthly precipitation in the metro area in inches. CAT and SCHOOL are binary variables indicating whether Dave was unavailable to work at TWS that month. CAT takes on a value of 0 if Dave was not vacationing that month, and 1 if he was; SCHOOL is 0 if Dave was not in school that month, and 1 if he was. EMPLOY represents the number of employees on the payroll that month. If there was an auction in town, the variable AUCTION is 1 (0 for no auction).

Dave has hired you to look at the data for him. He thinks the data can shed some light on two general areas:

1 Factors that influence fluctuations in revenues of TWS. Provide an overview of the data. Does it appear that Dave is right about his absenteeism hurting the business? What other factors influence fluctuations in revenues of TWS?

2 Future values of TWS revenues. Dave would like to try his hand at forecasting future values of TWS revenues. If he is successful at doing this, budgeting for employee hours, equipment, and chemicals will be much easier.

For the second task, construct a forecasting model of total sales for TWS. Use this model to predict total sales for the upcoming months of March and April.

Write a business memo to Dave that addresses the issues raised above. Without relying on any jargon, explain to Dave how he might use your forecasting model to predict revenues later in the year.

DATA SET

SALES	SALES2	TEMP	PREC	CAT	SCHOOL	EMPLOY	AUCTION
4214	262	59.0	4.26	0	0	2.0	0
5595	560	71.9	1.28	0	0	2.0	0
3274	175	74.2	2.19	1	0	2.0	0
6075	420	73.6	1.83	0	0	2.0	1
4354	259	62.3	9.90	0	0	2.0	0
3994	200	51.3	0.06	1	0	2.0	0
6256	335	42.8	0.47	0	0	2.0	1
4150	540	27.3	1.04	0	0	2.0	0
4277	23	33.5	1.14	0	0	2.0	0
3955	812	22.4	0.66	0	0	2.0	0
6986	1486	43.3	0.56	0	0	2.0	0
5940	939	51.1	1.00	0	0	2.0	0
2973	261	59.0	3.83	0	0	2.0	0
5687	333	65.4	2.04	0	0	1.5	0
4779	95	75.9	1.64	1	0	1.0	0
5321	367	71.7	1.28	0	0	1.0	0
4849	1085	62.5	1.55	0	0	1.0	0
6484	525	51.3	0.81	0	0	1.0	0
5569	89	42.8	0.15	1	0	1.0	0
3834	1245	27.3	0.81	0	0	2.0	1
5469	188	36.4	0.74	1	0	2.0	0
5863	580	33.3	0.55	0	0	2.0	0
6148	175	39.5	3.10	0	0	2.0	0
5578	353	49.1	1.01	0	0	2.0	0

(concludes on the next page)

CASE 26 DAVE'S TRUCK WASHING SERVICE II–3

SALES	SALES2	TEMP	PREC	CAT	SCHOOL	EMPLOY	AUCTION
8318	1377	56.6	1.51	0	0	2.0	0
7132	663	72.6	0.21	0	0	2.0	1
6888	95	70.8	3.57	0	0	2.0	0
8301	200	71.3	1.96	0	0	2.0	1
6838	432	66.9	1.46	0	0	2.0	0
6145	230	52.3	1.03	0	0	2.0	0
6470	325	44.0	1.28	1	0	2.0	0
4580	0	25.7	0.27	1	0	2.0	1
6555	100	27.9	0.76	0	0	2.5	0
6540	205	40.3	0.08	0	0	2.5	0
5874	230	42.8	0.76	0	0	2.5	0
5692	0	47.8	1.94	1	0	2.5	0
5922	475	60.3	2.43	1	0	2.5	0
5189	1100	69.4	2.20	1	0	2.5	0
5458	347	73.1	4.11	0	0	2.5	0
10165	2657	71.5	3.69	0	0	2.5	1
5829	631	62.5	0.79	0	1	2.5	0
4210	460	50.3	0.70	1	1	2.5	0
5066	784	34.7	2.67	0	1	2.5	1
6960	885	33.1	0.19	0	0	2.5	0
5484	378	31.8	1.19	0	1	2.5	0
5892	1020	40.1	0.10	0	1	2.5	0

CASE 27

HOUSING PRICES—I

You currently own a house in Eastville, Oregon, and are planning on moving sometime in the next year. A friend of yours has suggested that you try to sell the house on your own. The decision is a difficult one. Real estate commissions can be quite hefty, but brokers bring a great deal of expertise to the process of selling a house, particularly the ability to set a reasonable asking price. Real estate professionals can quantify a variety of unobservable or unmeasurable characteristics of a house, such as that shiny purple wallpaper in the upstairs bathroom and the overall desirability of the neighborhood.

To get an initial feel for the types of houses in your area, you have collected the data in the file named HOUSES. This file contains information such as the selling price of the house, square feet, numbers of bedrooms and bathrooms, and the presence of a fireplace for 108 homes sold in your neighborhood last winter. In the data set:

SQ_FT is the variable measuring the total square feet in the house. BEDS and BATHS are number of bedrooms and bathrooms, respectively. HEAT and STYLE are categorical variables. HEAT takes on the value of 0 for gas forced air heating and 1 for electric heat. STYLE is the architectural style of the home: 0 indicates a trilevel, 1 indicates a two-story house, and 2 indicates that the house is a ranch-styled home. GARAGE is the number of cars that can fit into the garage. We don't know whether it is an attached garage. AGE is the age of the home in years. FIRE and BASEMENT indicate the presence (1) or absence (0) of at least one fireplace or basement. PRICE is the selling price of the house in thousands of dollars and SCHOOL is the school district (0 = Eastville school district; 1 = Apple Valley school district). All else the same, Apple Valley is viewed as the preferred school district.

These data were used by Ellen L. Chilikas, David S. Abelson, and Brian R. Landry to price a house owned by one of the group members as part of a student project. The name of the suburb from which the data were taken has been changed.

Your task is to use these data to answer a couple of questions.

- What is the nature of the selling prices of these houses?
- What is the breakout of selling prices in terms of architectural style?
- How does the existence of a basement contribute to selling price?
- How does the existence of a fireplace add to the selling price?
- What other interesting or useful information do you see in this data set?

DATA SET

SQ_FT	BEDS	BATHS	HEAT	STYLE	GARAGE	BASEMENT	AGE	FIRE	PRICE	SCHOOL
1238	3	2	0	0	1	1	12	1	59.900	1
1707	3	2	1	0	2	0	13	1	64.000	0
1296	4	2	0	0	2	1	17	0	66.500	0
1320	3	2	0	0	2	1	11	1	66.500	0
1210	3	2	0	0	1	0	6	1	66.900	0
1296	3	2	0	0	2	1	17	1	68.000	0
1765	3	2	0	0	2	1	20	0	68.500	0
1725	4	3	0	0	2	1	12	0	69.000	0
1794	4	2	0	0	2	1	18	0	70.950	0
1294	3	2	0	0	2	0	13	1	71.000	0
1372	3	2	0	0	2	1	9	0	72.692	1
1162	3	2	0	0	1	0	8	1	72.801	0
1996	4	2	0	0	2	0	13	1	75.207	1
1764	4	2	1	0	2	1	13	1	76.000	0
1416	3	2	0	0	2	0	8	0	76.000	1
1730	4	2	0	0	2	0	15	1	77.500	0
1392	3	2	0	0	2	1	8	1	79.900	1
1664	3	3	0	0	2	0	11	1	79.900	0
1332	3	2	0	0	2	1	14	0	81.000	1
1752	3	3	0	0	2	0	18	1	82.800	0
2167	3	3	1	0	2	1	13	1	84.900	0
1664	3	2	0	0	2	0	9	1	85.000	0
1973	4	3	0	0	2	0	13	1	86.000	0
1384	3	2	0	0	2	0	5	1	89.280	0
1431	3	2	0	0	2	1	7	1	89.900	1
1950	5	3	0	0	2	0	13	1	90.000	0
1452	3	2	0	0	2	1	4	1	92.000	0
1829	3	2	0	0	2	0	10	1	92.439	1
1652	4	3	0	0	2	1	7	1	94.646	1
1516	3	2	0	0	2	1	10	1	97.293	1
1998	4	3	0	0	2	1	17	1	98.100	1
1984	4	3	0	0	2	1	9	1	98.149	0
1840	3	3	0	0	2	1	9	1	105.000	0
1823	4	3	0	0	2	1	3	1	110.000	1
2150	5	3	0	0	2	1	12	1	111.900	0
2096	3	3	0	0	2	1	9	1	113.000	1
2212	4	3	0	0	3	1	17	1	124.000	1

(continues on the following page)

CASE 27 HOUSING PRICES—I

SQ_FT	BEDS	BATHS	HEAT	STYLE	GARAGE	BASEMENT	AGE	FIRE	PRICE	SCHOOL
2375	4	3	0	0	2	1	11	1	133.000	1
2809	4	3	0	0	3	1	6	1	195.000	0
912	4	2	0	1	2	1	19	1	59.000	0
816	3	2	0	1	2	1	19	0	61.500	0
1008	5	2	0	1	2	1	17	0	63.500	0
912	4	2	0	1	2	1	16	1	66.950	1
1008	5	2	0	1	2	1	17	1	68.000	0
838	4	2	0	1	1	1	11	0	68.694	1
1008	4	3	0	1	2	1	19	1	69.000	0
1120	3	2	0	1	2	1	11	1	70.452	1
1242	4	3	0	1	2	1	12	0	75.000	0
912	4	2	0	1	2	1	18	1	76.900	1
1316	3	2	0	1	2	1	8	1	83.000	0
1490	3	2	0	1	2	1	13	1	88.000	0
1118	4	3	0	1	2	1	16	1	88.879	1
1278	3	2	0	1	2	1	11	1	89.347	0
1490	3	2	0	1	2	1	10	1	89.700	1
1418	4	2	0	1	2	1	14	1	90.000	1
1250	4	3	0	1	2	1	17	1	90.800	1
1881	4	3	0	1	2	1	9	1	91.500	1
1355	3	2	0	1	2	1	7	1	95.000	1
1518	3	2	0	1	2	1	22	0	98.000	1
1606	4	3	0	1	2	1	13	1	99.500	0
2124	3	2	0	1	2	1	14	1	107.000	0
2099	4	3	0	1	2	1	13	1	113.750	1
2232	5	3	0	1	2	1	9	1	125.000	1
1608	5	3	0	1	3	1	13	1	132.000	1
1320	3	2	0	2	2	1	13	1	76.000	1
1680	3	3	0	2	2	0	7	1	79.900	1
1434	3	3	0	2	2	1	12	1	79.900	1
1664	3	2	0	2	2	1	7	1	85.000	1
1379	3	2	0	2	2	1	8	1	87.158	1
1634	3	3	0	2	2	1	10	1	87.200	1
1400	3	3	0	2	2	0	1	1	91.500	1
1664	4	4	0	2	2	1	16	1	91.500	1
1548	3	3	0	2	2	1	8	1	92.500	1
2040	4	3	0	2	2	1	15	1	93.500	0
1792	4	3	0	2	2	1	9	1	94.000	1
1865	3	3	0	2	2	1	12	1	96.700	0
1833	3	3	0	2	2	1	7	1	97.400	1
1627	3	3	0	2	2	1	3	1	98.500	1
2198	4	3	0	2	2	1	12	1	99.000	0
1707	3	3	0	2	2	1	12	1	99.000	0
2096	3	3	0	2	2	1	7	1	102.400	1
2004	4	3	0	2	2	1	13	1	104.000	0
1818	4	3	0	2	2	1	19	1	107.000	1

(concludes on the following page)

CASE 27 HOUSING PRICES—I

SQ_FT	BEDS	BATHS	HEAT	STYLE	GARAGE	BASEMENT	AGE	FIRE	PRICE	SCHOOL
1935	3	3	0	2	2	1	11	1	107.900	1
2381	3	3	0	2	2	0	17	1	109.000	1
1977	3	3	0	2	2	1	9	1	110.000	1
2317	5	3	0	2	2	1	9	1	110.500	1
2072	4	3	0	2	2	1	11	1	112.000	1
2108	3	3	0	2	2	1	7	1	112.900	0
2060	4	4	0	2	2	1	9	1	113.600	1
2064	3	3	0	2	2	1	5	1	114.500	1
2148	5	4	0	2	2	1	8	1	114.700	1
2170	6	4	0	2	2	1	21	1	115.000	1
1984	4	3	1	2	2	1	11	1	121.000	1
2130	4	3	0	2	2	1	20	1	121.000	1
2049	4	3	0	2	2	1	19	1	125.000	1
2035	3	3	0	2	3	1	6	1	128.000	1
2264	3	3	0	2	3	1	7	1	130.000	1
2264	3	3	0	2	3	1	10	1	131.000	1
2362	4	3	0	2	2	1	6	1	132.000	0
2262	5	3	0	2	2	1	12	1	132.500	1
2353	4	3	0	2	2	1	5	1	132.500	0
2298	3	3	0	2	3	1	2	1	144.500	1
2282	4	3	0	2	2	1	9	1	150.577	1
2542	4	3	0	2	2	1	6	1	153.800	1
2380	4	3	0	2	2	1	6	1	155.000	1
2505	4	3	0	2	2	1	7	1	156.900	1
2804	4	3	0	2	3	1	1	1	192.000	1

CASE 28

HOUSING PRICES—II

You currently own a home in Eastville, Oregon, and want to put your house on the market. You're not in any particular hurry to get rid of the house, and would like to try selling it yourself. One way to determine a reasonable asking price of a house is to call one or more real estate agents and seek their advice. Another is to hire an appraiser—this approach would cost several hundred dollars. You've been wondering if there might be an easier and cheaper way to understand what determines selling prices in the area.

On your daily after-dinner stroll through the neighborhood, you've passed several houses on the market. At first, you hoped that you could get a feel for the market by simply studying these few houses. But the more you think about this problem, the more confused you get. You're pretty sure that bigger homes sell for more, but you don't know how *much* more. Also, it seems as if fireplaces would be a desirable amenity in Oregon; your house doesn't have a fireplace, and you're not sure how much to lower the asking price because of this. You've also heard that location is the most important thing in the real estate market, and you wonder if prices tend to vary by school district, one indication of the quality of the neighborhood.

A friend has collected some information for you about selling prices and other characteristics of houses sold within the last few months in your neighborhood. These include square feet, number of bedrooms and bathrooms, type of heating, and school district. This data set is shown below. At first, you decided that your problem was solved—you would only need to find a house just like yours in the data set, and use its price as your guess.

Of course, it's never that easy. Although you know the specific characteristics of your house, such as square feet, number of bedrooms and bathrooms, and so forth, *not*

These data were used by Ellen L. Chilikas, David S. Abelson, and Brian R. Landry to price a house owned by one of the group members as part of a student project. The name of the suburb from which the data were taken has been changed.

one of the houses in the data set is a perfect match for your own home. Some of the houses in the data set come *close* to yours, but every comparable house is different from yours in at least three aspects. You're wondering if there might not be another way to use the data to come up with an estimate.

The data are contained in the file named HOUSES, which includes information on 108 houses. In this data set, SQ_FT is the variable measuring the total square feet in the house. BEDS and BATHS are number of bedrooms and bathrooms, respectively. HEAT and STYLE are categorical variables. HEAT takes on the value of 0 for gas forced air heating and 1 for electric heat. STYLE is the architectural style of the home: 0 indicates a trilevel, 1 indicates a two-story house, and 2 indicates that the house is a ranch-styled home. GARAGE is the number of cars that might fit in the garage, although you don't know whether the garage is attached. AGE is the age of the house in years. FIRE indicates the presence (FIRE = 1) or absence (FIRE = 0) of a fireplace and BASEMENT indicates the presence (BASEMENT = 1) or absence (BASEMENT = 0) of a basement. PRICE is the selling price of the house in thousands of dollars and SCHOOL is the school district (0 = Eastville school district; 1 = Apple Valley school district). All else the same, Apple Valley is viewed as being the more desirable of the two school districts.

Use these data to uncover the important determinants of selling prices of houses in your neighborhood. Then prepare a description of how your findings might be used as a general method for estimating the selling price of *any* house in your neighborhood, such as yours.

DATA SET

SQ_FT	BEDS	BATHS	HEAT	STYLE	GARAGE	BASEMENT	AGE	FIRE	PRICE	SCHOOL
1238	3	2	0	0	1	1	12	1	59.900	1
1707	3	2	1	0	2	0	13	1	64.000	0
1296	4	2	0	0	2	1	17	0	66.500	0
1320	3	2	0	0	2	1	11	1	66.500	0
1210	3	2	0	0	1	0	6	1	66.900	0
1296	3	2	0	0	2	1	17	1	68.000	0
1765	3	2	0	0	2	1	20	0	68.500	0
1725	4	3	0	0	2	1	12	0	69.000	0
1794	4	2	0	0	2	1	18	0	70.950	0
1294	3	2	0	0	2	0	13	1	71.000	0
1372	3	2	0	0	2	1	9	0	72.692	1
1162	3	2	0	0	1	0	8	1	72.801	0
1996	4	2	0	0	2	0	13	1	75.207	1
1764	4	2	1	0	2	1	13	1	76.000	0
1416	3	2	0	0	2	0	8	0	76.000	1
1730	4	2	0	0	2	0	15	1	77.500	0
1392	3	2	0	0	2	1	8	1	79.900	1
1664	3	3	0	0	2	0	11	1	79.900	0
1332	3	2	0	0	2	1	14	0	81.000	1
1752	3	3	0	0	2	0	18	1	82.800	0
2167	3	3	1	0	2	1	13	1	84.900	0
1664	3	2	0	0	2	0	9	1	85.000	0

(continues on the next page)

CASE 28 HOUSING PRICES—II   II–11

SQ_FT	BEDS	BATHS	HEAT	STYLE	GARAGE	BASEMENT	AGE	FIRE	PRICE	SCHOOL
1973	4	3	0	0	2	0	13	1	86.000	0
1384	3	2	0	0	2	0	5	1	89.280	0
1431	3	2	0	0	2	1	7	1	89.900	1
1950	5	3	0	0	2	0	13	1	90.000	0
1452	3	2	0	0	2	1	4	1	92.000	0
1829	3	2	0	0	2	0	10	1	92.439	1
1652	4	3	0	0	2	1	7	1	94.646	1
1516	3	2	0	0	2	1	10	1	97.293	1
1998	4	3	0	0	2	1	17	1	98.100	1
1984	4	3	0	0	2	1	9	1	98.149	0
1840	3	3	0	0	2	1	9	1	105.000	0
1823	4	3	0	0	2	1	3	1	110.000	1
2150	5	3	0	0	2	1	12	1	111.900	0
2096	3	3	0	0	2	1	9	1	113.000	1
2212	4	3	0	0	3	1	17	1	124.000	1
2375	4	3	0	0	2	1	11	1	133.000	1
2809	4	3	0	0	3	1	6	1	195.000	0
912	4	2	0	1	2	1	19	1	59.000	0
816	3	2	0	1	2	1	19	0	61.500	0
1008	5	2	0	1	2	1	17	0	63.500	0
912	4	2	0	1	2	1	16	1	66.950	1
1008	5	2	0	1	2	1	17	1	68.000	0
838	4	2	0	1	1	1	11	0	68.694	1
1008	4	3	0	1	2	1	19	1	69.000	0
1120	3	2	0	1	2	1	11	1	70.452	1
1242	4	3	0	1	2	1	12	0	75.000	0
912	4	2	0	1	2	1	18	1	76.900	1
1316	3	2	0	1	2	1	8	1	83.000	0
1490	3	2	0	1	2	1	13	1	88.000	0
1118	4	3	0	1	2	1	16	1	88.879	1
1278	3	2	0	1	2	1	11	1	89.347	0
1490	3	2	0	1	2	1	10	1	89.700	1
1418	4	2	0	1	2	1	14	1	90.000	1
1250	4	3	0	1	2	1	17	1	90.800	1
1881	4	3	0	1	2	1	9	1	91.500	1
1355	3	2	0	1	2	1	7	1	95.000	1
1518	3	2	0	1	2	1	22	0	98.000	1
1606	4	3	0	1	2	1	13	1	99.500	0
2124	3	2	0	1	2	1	14	1	107.000	0
2099	4	3	0	1	2	1	13	1	113.750	1
2232	5	3	0	1	2	1	9	1	125.000	1
1608	5	3	0	1	3	1	13	1	132.000	1
1320	3	2	0	2	2	1	13	1	76.000	1
1680	3	3	0	2	2	0	7	1	79.900	1
1434	3	3	0	2	2	1	12	1	79.900	1
1664	3	2	0	2	2	1	7	1	85.000	1
1379	3	2	0	2	2	1	8	1	87.158	1

(concludes on the next page)

CASE 28 HOUSING PRICES—II

SQ_FT	BEDS	BATHS	HEAT	STYLE	GARAGE	BASEMENT	AGE	FIRE	PRICE	SCHOOL
1634	3	3	0	2	2	1	10	1	87.200	1
1400	3	3	0	2	2	0	1	1	91.500	1
1664	4	4	0	2	2	1	16	1	91.500	1
1548	3	3	0	2	2	1	8	1	92.500	1
2040	4	3	0	2	2	1	15	1	93.500	0
1792	4	3	0	2	2	1	9	1	94.000	1
1865	3	3	0	2	2	1	12	1	96.700	0
1833	3	3	0	2	2	1	7	1	97.400	1
1627	3	3	0	2	2	1	3	1	98.500	1
2198	4	3	0	2	2	1	12	1	99.000	0
1707	3	3	0	2	2	1	12	1	99.000	0
2096	3	3	0	2	2	1	7	1	102.400	1
2004	4	3	0	2	2	1	13	1	104.000	0
1818	4	3	0	2	2	1	19	1	107.000	1
1935	3	3	0	2	2	1	11	1	107.900	1
2381	3	3	0	2	2	0	17	1	109.000	1
1977	3	3	0	2	2	1	9	1	110.000	1
2317	5	3	0	2	2	1	9	1	110.500	1
2072	4	3	0	2	2	1	11	1	112.000	1
2108	3	3	0	2	2	1	7	1	112.900	0
2060	4	4	0	2	2	1	9	1	113.600	1
2064	3	3	0	2	2	1	5	1	114.500	1
2148	5	4	0	2	2	1	8	1	114.700	1
2170	6	4	0	2	2	1	21	1	115.000	1
1984	4	3	1	2	2	1	11	1	121.000	1
2130	4	3	0	2	2	1	20	1	121.000	1
2049	4	3	0	2	2	1	19	1	125.000	1
2035	3	3	0	2	3	1	6	1	128.000	1
2264	3	3	0	2	3	1	7	1	130.000	1
2264	3	3	0	2	3	1	10	1	131.000	1
2362	4	3	0	2	2	1	6	1	132.000	0
2262	5	3	0	2	2	1	12	1	132.500	1
2353	4	3	0	2	2	1	5	1	132.500	0
2298	3	3	0	2	3	1	2	1	144.500	1
2282	4	3	0	2	2	1	9	1	150.577	1
2542	4	3	0	2	2	1	6	1	153.800	1
2380	4	3	0	2	2	1	6	1	155.000	1
2505	4	3	0	2	2	1	7	1	156.900	1
2804	4	3	0	2	3	1	1	1	192.000	1

CASE 29

BOOKS

The retail book market has become very competitive. These days bookstores offer more than just books: magazines, international and domestic newspapers, notecards, stationery, maps, books on tape, reading lights, and gifts are just some of the items found on bookstore shelves. Large retailers such as Barnes & Noble are able to offer low prices and large inventories to their customers, much to the chagrin of local bookstore owners. The large chain bookstores have recently added coffee shops or delicatessens to entice buyers to stay and browse.

A client of the consulting firm for which you work is considering investment into the retail book sales industry, and has hired your firm to undertake a financial study. Your job is to examine the data in the file named BOOKS. This data set contains general financial characteristics of five retail bookstores (Standard Industrial Classification [SIC] code 5942). The firms were chosen to include different-sized business entities: two large corporations (Barnes & Noble and Borders), two medium size firms (Books-A-Million and Crown) and, for comparison, one of the smaller stores—Village Green Bookstore.

Barnes & Noble, in addition to selling hardcover and paperback books, publishes books for exclusive sale by Barnes & Noble outlets. The firm also operates a retail book mail-order business. Borders Group, Inc. is a holding company which owns and operates book "superstores" and smaller stores such as those found in shopping malls. Borders bookstores offer prerecorded music and videotapes, and claim to specialize in hard-to-find items. Borders is one of the largest firms in its SIC group, employing almost 23,000 people in 1997.

Books-A-Million is a relatively new corporation offering the full line of hardcover and softcover books and magazines. Books-A-Million also distributes books to wholesale clubs, supermarkets, department stores, and mass merchandisers. Crown Books

These data and company descriptions come from various company reports.

offers the usual range of books, audiotapes, videotapes, and newspapers, and also some computer software. Headquartered in Columbus, Ohio, Crown has recently experienced financial difficulties: net income for Crown was −$860,000 in 1997.

Village Bookstore, out of Rochester, New York, sells much more than books. According to corporate reports, Village offers "books, newspapers, magazines, periodicals, greeting cards, boxed stationery, writing papers, books-on-tape, selective video tape rentals and multimedia products, humorous and traditional gift items and food items such as fresh and packaged desserts, gourmet chocolates, coffees and ice creams as well as other food items; and operates educational retail systems which sell children's educational toys and products."

The data set contains the longest time-series available for each of these five firms. Incomplete financial information is indicated with an asterisk. Use these data to undertake a study of this industry. What do the historical trends have to say about these firms? Use an appropriate statistical method to characterize net operating assets of these firms. What factors influence net operating assets? How? Prepare a summary of these matters for presentation at next week's meeting with your client.

DEFINITIONS

Firm	1 = Barnes & Noble, Inc.
	2 = Books-A-Million, Inc.
	3 = Borders Group, Inc.
	4 = Crown Books Corp.
	5 = Village Green Bookstore, Inc.
Year	Year, 19XX
PPE	Property, plant, and equipment - Total (net), in millions of $
Assets	Current assets - Total, in millions of $
Liab	Current liabilities - Total, in millions of $
OpAssets	Net operating assets - Total, in millions of $

DATA SET

Firm	Year	PPE	Assets	Liab	OpAssets
1	91	84.101	306.168	332.338	57.933
1	92	92.638	336.898	364.789	64.747
1	93	135.985	415.416	300.739	250.662
1	94	190.187	546.342	363.939	372.590
1	95	243.635	613.872	457.896	399.611
1	96	319.880	848.188	621.688	546.380
1	97	434.774	866.955	654.263	647.466
2	91	5.338	30.812	*	*
2	92	6.157	34.758	*	*
2	93	10.353	57.239	24.455	43.137
2	94	15.599	99.333	28.261	86.671
2	95	35.864	125.843	29.112	132.595
2	96	49.253	140.003	40.151	149.105
2	97	63.147	168.755	75.650	156.252
3	94	208.200	521.900	439.700	290.400
3	95	244.200	816.200	558.200	502.200

(concludes on the following page)

Firm	Year	PPE	Assets	Liab	OpAssets
3	96	243.500	740.200	542.900	440.800
3	97	289.200	846.400	634.500	501.100
4	88	7.541	85.032	31.177	61.396
4	89	7.318	96.213	34.588	68.943
4	90	8.091	100.562	37.631	71.022
4	91	9.637	127.097	49.495	87.239
4	92	10.714	149.432	62.622	97.524
4	93	15.018	147.809	56.492	106.335
4	94	25.228	180.210	94.981	110.457
4	95	23.004	182.331	105.553	99.782
4	96	23.762	150.574	77.606	96.730
4	97	25.379	141.583	84.515	82.447
5	88	1.462	2.907	1.651	2.718
5	89	1.380	2.713	1.757	2.336
5	90	0.496	3.060	2.127	1.429
5	91	0.499	3.335	2.061	1.773
5	92	0.515	3.394	2.002	1.907
5	93	0.615	5.004	4.137	1.482
5	94	0.670	4.815	3.304	2.181
5	95	1.237	5.760	5.478	1.519
5	96	1.934	7.112	5.191	3.855
5	97	1.331	3.101	3.042	1.390

CASE **30**

ELISA PLATE UNIFORMITY

ELISAs, Inc., manufactures plates used by hospital laboratories to test for the presence of antibodies to viruses such as lupus or HIV. The specific test process, called the ELISA method, consists of coating the bottom of plastic wells (like little test tubes) with a substance that contains molecules with which the specific antibodies will react. Serums from the patient are placed in the well, and if they contain the specific antibodies, those antibodies will bind to the bottom of the well, too. The test produces a blue color if antibodies are present. The amount of blueness depends on the number of antibodies present. The degree of blueness is then measured by a machine that produces readings of "optical densities," the final value of interest, on a scale of 0.000 to 3.000. A low reading on this scale indicates low numbers of antibodies present.

The plastic wells are manufactured in lots of 500 plates. Each plate contains 96 wells, arranged in a pattern of 8 rows and 12 columns, as shown below:

Column number

	1	2	3	4	5	6	7	8	9	10	11	12
Injector 1 ->	O	O	O	O	O	O	O	O	O	O	O	O
Injector 2 ->	O	O	O	O	O	O	O	O	O	O	O	O
Injector 3 ->	O	O	O	O	O	O	O	O	O	O	O	O
Injector 4 ->	O	O	O	O	O	O	O	O	O	O	O	O
Injector 5 ->	O	O	O	O	O	O	O	O	O	O	O	O
Injector 6 ->	O	O	O	O	O	O	O	O	O	O	O	O
Injector 7 ->	O	O	O	O	O	O	O	O	O	O	O	O
Injector 8 ->	X	O	O	O	O	O	O	O	O	O	O	O

This case describes a real situation, though the name of the company has been changed. (The ELISA process is real.) Information and data about the situation are used here with permission. They were provided by a team of students, including Susilo Hertanto, Michael Tigges, and several others.

CASE 30 ELISA PLATE UNIFORMITY

The machine that fills the wells with the substance to which the antibodies bind has a single bank of eight injectors, which correspond to the rows in the diagram above. The machine first fills the eight wells in column 1 simultaneously: each cell is filled by a squirt of solution from the injector directly over it. The bank of injectors then moves to column 2 and fills those eight wells simultaneously, and so forth. After filling the wells in column 12, the entire bank of injectors returns to the positions over column 1, in preparation for the next plate. Differences in the resulting wells, then, may be attributed to variations between the columns and/or variations between the injectors, as well as to variations in the other parts of the total measurement process.

In principle, if the wells are filled with identical serums (from the same patient) and if the resulting optical densities are measured by using the approved method, the results should not vary. In practice, of course, they do.

ELISAs, Inc., has provided the data below, which consists of the readings for 95 observations on each of five plates taken sequentially from a single lot. The same patient's serums were used throughout. The specific plates tested were designed to detect antibodies of the disease lupus. Note that for each plate, the well in row 8, column 1 (marked with an X in the diagram), is used as a control. Its value in the data below is "missing." ELISAs, Inc., wants to know how well (i.e., how uniformly) its process works and to understand any ways in which it might be improved.

Based on the data below (also available in file ELISA), how would you advise ELISAs, Inc.?

DEFINITIONS

PLATE	Plate number (1 through 5)
ROW	Row number (injector number, 1 through 8)
COL	Column number (1 through 12)
OPTDEN	Optical density (0.000 through 3.000)

DATA SET

OBS	PLATE	ROW	COL	OPTDEN	OBS	PLATE	ROW	COL	OPTDEN
1	1	1	1	1.421	18	1	2	3	1.334
2	1	2	1	1.446	19	1	3	3	1.287
3	1	3	1	1.614	20	1	4	3	1.084
4	1	4	1	1.578	21	1	5	3	1.199
5	1	5	1	1.452	22	1	6	3	1.120
6	1	6	1	1.790	23	1	7	3	1.532
7	1	7	1	1.991	24	1	8	3	1.389
8	1	8	1	*	25	1	1	4	1.466
9	1	1	2	1.266	26	1	2	4	1.298
10	1	2	2	1.270	27	1	3	4	1.288
11	1	3	2	1.106	28	1	4	4	1.165
12	1	4	2	1.159	29	1	5	4	1.293
13	1	5	2	1.141	30	1	6	4	1.392
14	1	6	2	1.002	31	1	7	4	1.523
15	1	7	2	1.207	32	1	8	4	1.657
16	1	8	2	1.259	33	1	1	5	1.454
17	1	1	3	1.804	34	1	2	5	1.356

(continues on the following page)

OBS	PLATE	ROW	COL	OPTDEN	OBS	PLATE	ROW	COL	OPTDEN
35	1	3	5	1.522	83	1	3	11	1.705
36	1	4	5	1.392	84	1	4	11	1.586
37	1	5	5	1.483	85	1	5	11	1.531
38	1	6	5	1.633	86	1	6	11	1.778
39	1	7	5	1.744	87	1	7	11	1.675
40	1	8	5	1.550	88	1	8	11	1.889
41	1	1	6	1.578	89	1	1	12	1.742
42	1	2	6	1.370	90	1	2	12	1.806
43	1	3	6	1.788	91	1	3	12	1.803
44	1	4	6	1.442	92	1	4	12	1.553
45	1	5	6	1.383	93	1	5	12	1.282
46	1	6	6	1.786	94	1	6	12	1.568
47	1	7	6	1.844	95	1	7	12	1.505
48	1	8	6	1.778	96	1	8	12	1.752
49	1	1	7	1.503	97	2	1	1	1.610
50	1	2	7	1.604	98	2	2	1	1.701
51	1	3	7	1.771	99	2	3	1	1.946
52	1	4	7	1.500	100	2	4	1	1.815
53	1	5	7	1.586	101	2	5	1	1.838
54	1	6	7	1.803	102	2	6	1	2.061
55	1	7	7	1.679	103	2	7	1	1.665
56	1	8	7	2.013	104	2	8	1	*
57	1	1	8	1.645	105	2	1	2	1.479
58	1	2	8	1.695	106	2	2	2	1.340
59	1	3	8	1.891	107	2	3	2	1.207
60	1	4	8	1.628	108	2	4	2	1.218
61	1	5	8	1.840	109	2	5	2	1.160
62	1	6	8	1.611	110	2	6	2	1.093
63	1	7	8	1.847	111	2	7	2	1.112
64	1	8	8	1.802	112	2	8	2	1.272
65	1	1	9	1.717	113	2	1	3	1.243
66	1	2	9	1.814	114	2	2	3	1.198
67	1	3	9	1.875	115	2	3	3	1.241
68	1	4	9	1.778	116	2	4	3	1.111
69	1	5	9	1.424	117	2	5	3	1.210
70	1	6	9	2.001	118	2	6	3	1.239
71	1	7	9	1.885	119	2	7	3	1.218
72	1	8	9	1.652	120	2	8	3	1.245
73	1	1	10	1.714	121	2	1	4	1.197
74	1	2	10	1.874	122	2	2	4	1.125
75	1	3	10	1.957	123	2	3	4	1.246
76	1	4	10	1.884	124	2	4	4	1.210
77	1	5	10	1.762	125	2	5	4	1.074
78	1	6	10	1.873	126	2	6	4	1.196
79	1	7	10	1.753	127	2	7	4	1.267
80	1	8	10	1.708	128	2	8	4	1.303
81	1	1	11	1.569	129	2	1	5	1.020
82	1	2	11	1.799	130	2	2	5	1.083

(continues on the following page)

OBS	PLATE	ROW	COL	OPTDEN	OBS	PLATE	ROW	COL	OPTDEN
131	2	3	5	1.087	179	2	3	11	1.184
132	2	4	5	1.148	180	2	4	11	1.400
133	2	5	5	1.190	181	2	5	11	1.220
134	2	6	5	1.184	182	2	6	11	1.276
135	2	7	5	1.298	183	2	7	11	1.350
136	2	8	5	1.487	184	2	8	11	1.476
137	2	1	6	1.148	185	2	1	12	1.391
138	2	2	6	1.133	186	2	2	12	1.511
139	2	3	6	1.069	187	2	3	12	1.233
140	2	4	6	1.137	188	2	4	12	1.284
141	2	5	6	1.116	189	2	5	12	1.220
142	2	6	6	1.190	190	2	6	12	1.230
143	2	7	6	1.368	191	2	7	12	1.363
144	2	8	6	1.416	192	2	8	12	1.366
145	2	1	7	1.137	193	3	1	1	1.531
146	2	2	7	1.223	194	3	2	1	1.946
147	2	3	7	1.226	195	3	3	1	1.662
148	2	4	7	1.282	196	3	4	1	1.799
149	2	5	7	1.040	197	3	5	1	1.596
150	2	6	7	1.318	198	3	6	1	1.504
151	2	7	7	1.251	199	3	7	1	1.415
152	2	8	7	1.625	200	3	8	1	*
153	2	1	8	1.176	201	3	1	2	1.575
154	2	2	8	1.327	202	3	2	2	1.675
155	2	3	8	1.207	203	3	3	2	1.298
156	2	4	8	1.260	204	3	4	2	1.179
157	2	5	8	1.112	205	3	5	2	1.253
158	2	6	8	1.332	206	3	6	2	1.221
159	2	7	8	1.419	207	3	7	2	1.019
160	2	8	8	1.547	208	3	8	2	0.943
161	2	1	9	1.209	209	3	1	3	1.915
162	2	2	9	1.302	210	3	2	3	1.582
163	2	3	9	1.309	211	3	3	3	1.538
164	2	4	9	1.543	212	3	4	3	1.577
165	2	5	9	1.259	213	3	5	3	1.444
166	2	6	9	1.423	214	3	6	3	1.276
167	2	7	9	1.448	215	3	7	3	1.467
168	2	8	9	1.578	216	3	8	3	1.398
169	2	1	10	1.255	217	3	1	4	1.199
170	2	2	10	1.319	218	3	2	4	1.407
171	2	3	10	1.269	219	3	3	4	1.267
172	2	4	10	1.356	220	3	4	4	1.170
173	2	5	10	1.195	221	3	5	4	1.257
174	2	6	10	1.305	222	3	6	4	1.119
175	2	7	10	1.434	223	3	7	4	1.252
176	2	8	10	1.439	224	3	8	4	1.290
177	2	1	11	1.249	225	3	1	5	1.213
178	2	2	11	1.292	226	3	2	5	1.310

(continues on the following page)

CASE 30 ELISA PLATE UNIFORMITY II–21

OBS	PLATE	ROW	COL	OPTDEN	OBS	PLATE	ROW	COL	OPTDEN
227	3	3	5	1.425	275	3	3	11	1.668
228	3	4	5	1.327	276	3	4	11	1.584
229	3	5	5	1.268	277	3	5	11	1.175
230	3	6	5	1.279	278	3	6	11	1.446
231	3	7	5	1.344	279	3	7	11	1.684
232	3	8	5	1.272	280	3	8	11	1.284
233	3	1	6	1.152	281	3	1	12	2.057
234	3	2	6	1.427	282	3	2	12	1.801
235	3	3	6	1.465	283	3	3	12	1.562
236	3	4	6	1.509	284	3	4	12	2.054
237	3	5	6	1.381	285	3	5	12	1.636
238	3	6	6	1.338	286	3	6	12	1.769
239	3	7	6	1.435	287	3	7	12	1.759
240	3	8	6	1.446	288	3	8	12	1.922
241	3	1	7	1.437	289	4	1	1	1.425
242	3	2	7	1.574	290	4	2	1	1.867
243	3	3	7	1.503	291	4	3	1	1.625
244	3	4	7	1.722	292	4	4	1	1.460
245	3	5	7	1.506	293	4	5	1	1.491
246	3	6	7	1.394	294	4	6	1	1.222
247	3	7	7	1.446	295	4	7	1	1.371
248	3	8	7	1.695	296	4	8	1	*
249	3	1	8	1.493	297	4	1	2	1.416
250	3	2	8	1.535	298	4	2	2	1.192
251	3	3	8	1.457	299	4	3	2	1.246
252	3	4	8	1.592	300	4	4	2	1.130
253	3	5	8	1.565	301	4	5	2	1.283
254	3	6	8	1.476	302	4	6	2	1.116
255	3	7	8	1.423	303	4	7	2	1.156
256	3	8	8	1.552	304	4	8	2	1.295
257	3	1	9	1.514	305	4	1	3	1.185
258	3	2	9	1.554	306	4	2	3	1.149
259	3	3	9	1.527	307	4	3	3	1.174
260	3	4	9	1.761	308	4	4	3	1.198
261	3	5	9	1.553	309	4	5	3	1.163
262	3	6	9	1.618	310	4	6	3	1.096
263	3	7	9	1.623	311	4	7	3	1.268
264	3	8	9	1.503	312	4	8	3	1.278
265	3	1	10	1.341	313	4	1	4	1.107
266	3	2	10	1.591	314	4	2	4	0.968
267	3	3	10	1.650	315	4	3	4	1.306
268	3	4	10	1.517	316	4	4	4	1.364
269	3	5	10	1.435	317	4	5	4	1.200
270	3	6	10	1.607	318	4	6	4	1.175
271	3	7	10	1.338	319	4	7	4	1.304
272	3	8	10	1.636	320	4	8	4	1.286
273	3	1	11	1.449	321	4	1	5	1.333
274	3	2	11	1.581	322	4	2	5	1.287

(continues on the following page)

II-22 CASE 30 ELISA PLATE UNIFORMITY

OBS	PLATE	ROW	COL	OPTDEN	OBS	PLATE	ROW	COL	OPTDEN
323	4	3	5	1.198	371	4	3	11	1.530
324	4	4	5	1.379	372	4	4	11	1.432
325	4	5	5	1.442	373	4	5	11	1.491
326	4	6	5	1.524	374	4	6	11	1.418
327	4	7	5	1.458	375	4	7	11	1.646
328	4	8	5	1.375	376	4	8	11	1.536
329	4	1	6	1.480	377	4	1	12	1.793
330	4	2	6	1.340	378	4	2	12	1.742
331	4	3	6	1.280	379	4	3	12	1.970
332	4	4	6	1.508	380	4	4	12	1.591
333	4	5	6	1.399	381	4	5	12	1.992
334	4	6	6	1.513	382	4	6	12	1.815
335	4	7	6	1.608	383	4	7	12	1.692
336	4	8	6	1.001	384	4	8	12	1.380
337	4	1	7	1.296	385	5	1	1	1.794
338	4	2	7	1.279	386	5	2	1	2.190
339	4	3	7	1.426	387	5	3	1	1.981
340	4	4	7	1.338	388	5	4	1	2.348
341	4	5	7	1.312	389	5	5	1	2.011
342	4	6	7	1.212	390	5	6	1	2.024
343	4	7	7	1.477	391	5	7	1	2.000
344	4	8	7	1.423	392	5	8	1	*
345	4	1	8	1.374	393	5	1	2	1.913
346	4	2	8	1.693	394	5	2	2	1.012
347	4	3	8	1.504	395	5	3	2	1.683
348	4	4	8	1.414	396	5	4	2	1.908
349	4	5	8	1.523	397	5	5	2	1.690
350	4	6	8	1.378	398	5	6	2	1.610
351	4	7	8	1.680	399	5	7	2	1.656
352	4	8	8	1.293	400	5	8	2	1.684
353	4	1	9	1.596	401	5	1	3	1.669
354	4	2	9	1.646	402	5	2	3	1.571
355	4	3	9	1.647	403	5	3	3	1.751
356	4	4	9	1.728	404	5	4	3	1.689
357	4	5	9	1.601	405	5	5	3	1.538
358	4	6	9	1.654	406	5	6	3	1.356
359	4	7	9	1.689	407	5	7	3	1.597
360	4	8	9	1.369	408	5	8	3	1.796
361	4	1	10	1.384	409	5	1	4	1.321
362	4	2	10	1.723	410	5	2	4	1.712
363	4	3	10	1.569	411	5	3	4	1.821
364	4	4	10	1.654	412	5	4	4	1.728
365	4	5	10	1.599	413	5	5	4	1.615
366	4	6	10	1.607	414	5	6	4	1.426
367	4	7	10	1.597	415	5	7	4	1.610
368	4	8	10	1.611	416	5	8	4	1.751
369	4	1	11	1.592	417	5	1	5	1.709
370	4	2	11	1.408	418	5	2	5	1.708

(concludes on the following page)

OBS	PLATE	ROW	COL	OPTDEN	OBS	PLATE	ROW	COL	OPTDEN
419	5	3	5	1.465	450	5	2	9	1.662
420	5	4	5	1.781	451	5	3	9	1.729
421	5	5	5	1.524	452	5	4	9	1.992
422	5	6	5	1.773	453	5	5	9	1.877
423	5	7	5	1.833	454	5	6	9	1.936
424	5	8	5	1.707	455	5	7	9	2.109
425	5	1	6	1.366	456	5	8	9	2.208
426	5	2	6	1.654	457	5	1	10	1.210
427	5	3	6	1.402	458	5	2	10	1.673
428	5	4	6	1.642	459	5	3	10	1.668
429	5	5	6	1.540	460	5	4	10	1.911
430	5	6	6	1.680	461	5	5	10	1.803
431	5	7	6	1.904	462	5	6	10	2.079
432	5	8	6	1.980	463	5	7	10	1.796
433	5	1	7	1.192	464	5	8	10	1.951
434	5	2	7	1.765	465	5	1	11	1.832
435	5	3	7	1.765	466	5	2	11	1.837
436	5	4	7	1.932	467	5	3	11	1.797
437	5	5	7	1.696	468	5	4	11	1.855
438	5	6	7	1.894	469	5	5	11	1.824
439	5	7	7	2.098	470	5	6	11	1.832
440	5	8	7	2.107	471	5	7	11	2.022
441	5	1	8	1.148	472	5	8	11	1.722
442	5	2	8	1.891	473	5	1	12	1.370
443	5	3	8	1.775	474	5	2	12	2.034
444	5	4	8	2.028	475	5	3	12	2.246
445	5	5	8	1.779	476	5	4	12	1.926
446	5	6	8	1.976	477	5	5	12	1.835
447	5	7	8	2.321	478	5	6	12	2.253
448	5	8	8	2.041	479	5	7	12	2.314
449	5	1	9	1.232	480	5	8	12	2.281

CASE 31

VIOLENCE ON TV—I

As the general manager of KTDS, the NBC affiliate in Tidusville, Oklahoma, Chris has a range of responsibilities, including programming, personnel, advertising, and public relations. His least favorite activity is responding to customer complaints. Unfortunately, there's been an unusually large number of complaints in the past few months from viewers and advertisers alike.

Most of the recent comments are objections to the level of violence on KTDS programs. Chris is sensitive to this issue because he has observed a gradual increase in violence on TV over the past 20 years. Chris really prefers the old-time movies in which dirty deeds were neatly sanitized and violent crimes occurred behind the scenes. He is sympathetic to the recent callers.

Nonetheless, he's in a tight spot. Chris knows that small, vocal groups do not necessarily represent the population at large. People who feel strongly about an issue are likely to speak out, while those who are content tend to remain silent. While the recent callers have denounced the level of violence on KTDS shows, Chris knows he must understand and serve *all* of the KTDS viewers.

Chris has a suspicion about the source of the recent calls. Four months ago, a flamboyant politician announced his candidacy for mayor. This candidate has received a great deal of air time on the local news due, in part, to his impassioned outbursts. Some people love him, others despise him, but almost everyone tunes in to the evening news in hopes of catching the latest controversy. One continuing theme of his platform is violence in America in general and violence on KTDS in particular. Over the past four months, the candidate has suggested that those opposed to violence in the media "let

---

This case is based on an idea of Ann Lee Bailey, an MBA student at the University of Colorado at Denver.

their voices be heard." Chris suspects that this fellow has inspired a large portion of the recent complaints to KTDS.

The situation has become so volatile that Chris is convinced he needs to take some kind of action. To sort out the hype from any real concerns of his constituents, Chris has commissioned you to help him get some structured feedback. He would like to undertake a survey (and has the staff to do the legwork), but he would like advice on the specifics. In particular, he needs to know:

- What questions should be asked, and what is the best way to ask them?
- How should survey participants be chosen and how many should be chosen?
- What sort of confidentiality guarantees should be made to participants, if any?
- What is the best way to contact potential participants?
- Should the station offer some kind of free gift to induce higher participation rates? If so, what?
- Should one survey be used to address the concerns of the viewers and another survey to appease the station's advertisers? If so, how should the surveys differ?

Prepare a report for Chris that addresses these and any other issues you think he should be aware of. Include in your report a copy of the questionnaire.

CASE 32

OVERDUE BILLS

Quick Stab Collection Agency (QSCA) collects bills in an eastern town. The company specializes in small accounts and avoids risky collections, such as those in which the debtor tends to be chronically late in payments or is known to be hostile.

The business can be very profitable. OSCA buys the rights to collect debts from their original owners at a substantial discount. For example, QSCA might pay $10 for the right to collect a $60 debt. QCSA takes the risk of not collecting the debt *at all,* of course, but often a single official-looking letter yields full or nearly full payment, particularly for small debts.

Profitability at QSCA depends critically on the number of days to collect the payment and on the size of the bill, as well as on the discount rate offered.

A random sample of accounts closed out during the months of January through June yielded the data set below (and in file OVERDUE). Write a brief memo to QSCA management advising them on the relationship, if any, between size of bill and number of days to collect. In this data set, the variable DAYS is the number of days to collect the payment, and BILL is the amount of the overdue bill in dollars, while TYPE =1 for residential accounts and 0 for commercial accounts.

CASE 32 OVERDUE BILLS

DATA SET

DAYS	BILL	TYPE	DAYS	BILL	TYPE	DAYS	BILL	TYPE
41	215	1	52	302	1	83	98	0
60	205	0	60	197	0	55	225	0
86	79	0	47	288	0	69	150	0
81	97	0	26	150	1	48	273	1
37	201	1	71	158	0	25	146	1
90	50	0	59	215	0	60	210	0
94	46	0	31	158	1	42	210	1
83	95	0	30	149	1	36	205	1
84	100	0	70	154	0	50	302	0
79	140	0	34	180	1	68	187	0
47	299	0	38	205	1	22	95	1
33	187	1	42	220	1	11	46	1
47	264	1	29	162	1	44	301	0
69	180	0	83	97	0	47	289	0
19	97	1	50	311	1	19	98	1
36	179	1	49	250	0	67	199	0
30	154	1	25	153	1	73	149	0
39	310	0	16	80	1	91	70	0
63	205	0	43	225	1	82	90	0
17	110	1	51	310	1	63	211	0
85	75	0	71	179	0	74	153	0
21	100	1	74	150	0	24	150	1
49	301	1	67	201	0	92	80	0
83	95	0	22	97	1	65	146	0
13	75	1	53	273	0	99	60	0
16	79	1	5	90	1	47	288	1
53	240	0	57	220	0	51	264	0
40	197	1	10	50	1	39	211	1
47	311	0	80	110	0	27	140	1
48	299	1	47	289	1	44	250	1
70	162	0	15	70	1	35	199	1
43	240	1	11	60	1	6	95	1

CASE **33**

THE BRAZING OPERATION

In one of the phases in manufacturing a precision flow meter, a vacuum brazing process is used to attach hardware to the flow meter tube. The process consists of placing the "fixtured" assembly in a sealed vacuum furnace, pumping a hard vacuum on the furnace, slowly heating the assemblies to approximately 2050 degrees Fahrenheit, and then rapidly cooling them down to the ambient temperature.

The company wants the assembly to have a metallurgical structure that resists corrosion. One aspect of such a structure is a good grain size, a measure of the size of the actual crystals in the crystalline structure of the flow wall. The grain size is measured by a standard specification called the grain count, a numerical reading between 1 and 10. The larger the grain count, the finer the crystalline structure (and the more corrosion resistant).

The following data (and in file BRAZING) were gathered from process control charts. The run numbers are not consecutive because a variety of operations, not just the brazing operation, are performed in the furnace.

Based on the data beginning on the next page, what recommendations do you make concerning the process?

DEFINITIONS

RUN	Run number
TEMPAVE	Average maximum temperature (degrees Fahrenheit)
TIMEAVE	Average time at that temperature (minutes)
VACLEVEL	Vacuum level (in torr units; higher numbers indicate a poorer vacuum)
COOLRATE	Cooling rate, in degrees Fahrenheit per minute
GRNCOUNT	Grain count, in standard units, computed from a sample of the resulting product

David R. Berge provided the data on which this case is based and described the manufacturing situation.

DATA SET

RUN	TEMPAVE	TIMEAVE	COOLRATE	VACLEVEL	GRNCOUNT
326	2045.9	12.4	342	2.6	6.6
327	2046.0	12.6	352	6.0	6.4
332	2046.1	12.4	331	5.0	7.4
335	2047.0	12.4	304	6.1	8.0
336	2044.5	12.6	376	7.5	6.4
337	2047.4	12.2	347	5.7	6.1
341	2046.6	12.6	391	4.7	5.8
346	2047.5	12.7	354	5.5	6.0
348	2046.3	12.2	310	5.0	7.7
349	2046.1	12.4	345	6.0	6.6
351	2046.6	12.3	305	5.0	8.1
353	2046.8	12.4	349	2.9	6.4
355	2047.2	12.3	311	6.2	7.6
358	2046.0	12.8	363	2.3	6.2
361	2045.2	12.7	354	2.2	6.8
365	2042.9	12.4	346	5.0	7.4
367	2048.6	12.3	302	3.7	7.2
369	2044.5	12.8	361	3.7	6.9
371	2046.6	12.4	326	5.0	7.2
373	2050.9	12.6	389	3.4	4.6
375	2050.8	12.7	390	6.5	3.8
377	2050.1	12.3	336	5.0	5.9
381	2048.3	12.4	306	3.7	7.0
385	2049.3	12.8	353	1.7	5.9
389	2048.3	12.6	357	4.9	5.9
391	2049.3	12.3	310	4.9	6.9
394	2049.6	12.6	363	2.0	5.1
397	2050.9	12.3	330	6.4	6.1
399	2050.5	12.6	351	2.2	5.0
403	2049.0	12.5	392	5.0	4.9
407	2048.0	12.4	337	3.4	6.6
412	2050.2	12.2	326	3.0	6.5
415	2050.1	12.7	363	4.4	5.1
417	2049.0	12.6	391	3.9	4.2
419	2048.7	12.7	385	5.0	4.9
421	2048.2	12.3	347	3.9	6.1
425	2048.4	12.6	372	7.4	6.1
427	2050.0	12.6	398	2.7	3.9
429	2049.6	12.3	311	6.0	6.4
430	2051.3	12.7	352	3.3	4.9
433	2046.8	12.6	350	7.0	6.1
437	2046.4	12.4	342	3.2	6.8
440	2048.3	12.4	338	6.3	6.4
449	2052.7	12.6	354	3.2	4.3
453	2050.2	12.7	355	6.7	5.1
456	2048.2	12.3	311	7.9	7.0

(concludes on the next page)

RUN	TEMPAVE	TIMEAVE	COOLRATE	VACLEVEL	GRNCOUNT
458	2050.1	12.4	315	4.4	5.8
463	2050.1	12.5	348	4.6	5.5
471	2050.1	12.6	352	7.0	5.8
473	2050.3	12.8	360	5.4	5.2
477	2049.7	12.7	353	7.0	5.9
481	2047.2	12.6	374	4.8	6.2
486	2048.8	12.2	347	6.9	6.3

CASE 34

LEADING INDICATORS

Decision makers need reliable forecasts of key business activities to efficiently run an organization. Failure to anticipate future demand for a product can lead to foregone revenue, customer dissatisfaction, or lost market share. On the other hand, overestimation of future demand may lead to surplus inventory, with associated costs. Accurate projections of *all* costs of doing business, from personnel to capital, are needed to prepare budgets for an upcoming fiscal year. Although predictions about future business needs rarely turn out to be perfectly reliable, all firms do some kind of forecasting to assist in planning.

Some businesses rely on leading indicators for preparation of forecasts. Leading indicators are *current* levels of economic activity believed to foreshadow the direction and magnitude of *future* activities. For instance, since durable goods such as refrigerators and cars are time-consuming and costly to produce, manufacturers are more likely to step up production only when future demand is expected to be strong. Demand for large-ticket items seems to depend on consumers' confidence in their economic future, since consumers can postpone purchases of durable goods when times are hard and expected to remain so. Consumer confidence is one of the many leading indicators watched closely by business executives.

Inventories, wholesale trade, shipments of durable goods, housing starts, and the producer price index are other important leading indicators. Economic indicators are collected and published by organizations such as the Federal Reserve Board, the Bureau of the Census, and Dismal Sciences. See, for instance, www.dismal.com or www.census.gov/briefrm/esbr/www/brief.html.

Leading indicators can help prepare company forecasts if the firm can isolate which indicators apply to their industry. For instance, we might expect that mortgage applications and the interest rate on mortgages might foreshadow future activities in the construction industry.

CASE 34 LEADING INDICATORS

Can you prepare a model to forecast the need for new orders for materials and supplies in the construction industry using the leading indicators available in the data in the file named LEADING? The definitions below include citations from the World Wide Web that will assist you in updating and revising the data set. Make use of explanatory material about these data on the web if necessary.

DEFINITIONS

Yr.Month	Year and month of the time series; 91.01 is January 1991, and so forth
Pop	U.S. resident population in thousands (source: U.S. Census Bureau, *Current Population Reports* www.census.gov/main/www/subjects.html)
PermitSA	Building permits (seasonally adjusted annual rate in thousands) for new, privately owned housing units; the data are based on 17,000 permit-issuing places through 1993, and 19,000 places beginning in 1994 (source: U.S. Census Bureau, *Housing Units Authorized by Building Permits,* C 40 www.census.gov/const/www/index.html)
PermitNA	Building permits (unadjusted monthly estimates in thousands) for new, privately owned housing units; the data are based on 17,000 permit-issuing places through 1993, and 19,000 places beginning in 1994 (source: U.S. Census Bureau, *Housing Units Authorized by Building Permits,* C 40 www.census.gov/const/www/index.html)
Const_SA	New orders for construction materials and supplies in the U.S., in millions of dollars, seasonally adjusted (source: U.S. Census Bureau, *Current Industrial Reports* www.census.gov/ftp/pub/indicator/www/m3/index.htm)
Const_NA	New orders for construction materials and supplies in the U.S., in millions of dollars, not seasonally adjusted (source: U.S. Census Bureau, *Current Industrial Reports* www.census.gov/ftp/pub/indicator/www/m3/index.htm)
MortRate	Interest rate on 30-year, fixed rate, conventional mortgage loans (source: Federal Home Loan Mortgage Corporation www.bog.frb.fed.us)

DATA SET

Yr.Month	Pop	PermitSA	Const_SA	Const_NA	PermitNA	MortRate
91.01	250660	786	14043	12320	51.5	9.64
91.02	250848	853	13695	13131	54.1	9.37
91.03	251028	911	13523	13826	78.7	9.50
91.04	251246	916	14199	14333	90.9	9.49
91.05	251524	991	14385	14735	98.3	9.47
91.06	251861	964	14389	15685	93.2	9.62
91.07	252106	973	15298	14845	91.3	9.58
91.08	252379	944	14566	15262	84.9	9.24

(continues on the following page)

Yr.Month	Pop	PermitSA	Const_SA	Const_NA	PermitNA	MortRate
91.09	252661	974	14645	15637	80.6	9.01
91.10	252932	991	14761	15445	89.8	8.86
91.11	253177	984	14553	14173	67.1	8.71
91.12	253385	1061	14457	13229	68.4	8.50
92.01	253589	1077	14790	12964	67.9	8.43
92.02	253780	1146	14653	14132	72.6	8.76
92.03	253993	1082	15085	15384	97.7	8.94
92.04	254238	1054	15305	15462	103.9	8.85
92.05	254488	1056	15185	15582	97.1	8.67
92.06	254749	1057	15860	17198	109.2	8.51
92.07	255011	1089	15102	14611	101.1	8.13
92.08	255304	1075	15275	16052	93.0	7.98
92.09	255577	1114	15864	16913	99.2	7.92
92.10	255843	1132	15649	16351	97.1	8.09
92.11	256082	1118	15523	15164	76.4	8.31
92.12	256297	1176	15961	14572	79.5	8.22
93.01	256512	1177	15872	13926	67.1	8.02
93.02	256700	1148	16766	16290	73.1	7.68
93.03	256889	1056	16307	16628	99.1	7.50
93.04	257100	1104	16248	16372	107.3	7.47
93.05	257318	1112	16566	17005	101.7	7.47
93.06	257551	1130	16284	17676	117.8	7.42
93.07	257795	1174	16079	15521	104.5	7.21
93.08	258056	1230	16305	17214	111.9	7.11
93.09	258307	1251	16639	17722	111.5	6.92
93.10	258555	1287	16660	17428	107.0	6.83
93.11	258766	1357	17185	16858	98.1	7.16
93.12	258965	1461	17426	15823	100.1	7.17
94.01	259159	1392	17231	15180	80.6	7.06
94.02	259312	1279	17421	16994	81.9	7.15
94.03	259494	1331	18476	18745	126.3	7.68
94.04	259723	1377	17985	18016	128.0	8.32
94.05	259918	1383	17971	18444	132.0	8.60
94.06	260140	1336	18201	19589	138.8	8.40
94.07	260372	1347	18143	17424	114.8	8.61
94.08	260610	1386	18342	19355	131.5	8.51
94.09	260853	1426	18384	19515	127.2	8.64
94.10	261087	1401	18747	19596	117.0	8.93
94.11	261297	1358	19049	18766	100.5	9.17
94.12	261496	1420	19350	17681	94.2	9.20
95.01	261687	1293	19750	17560	79.1	9.15
95.02	261838	1282	19474	18948	81.3	8.83
95.03	262024	1235	19030	19292	113.4	8.46
95.04	262251	1243	18548	18016	110.2	8.32
95.05	262451	1243	18807	18444	122.7	7.96
95.06	262661	1275	18529	19589	129.3	7.57
95.07	262890	1355	18902	17424	116.4	7.61

(concludes on the following page)

CASE 34 LEADING INDICATORS

Yr.Month	Pop	PermitSA	Const_SA	Const_NA	PermitNA	MortRate
95.08	263123	1368	19382	19355	132.3	7.86
95.09	263366	1405	19105	19515	121.6	7.64
95.10	263609	1384	19169	19596	121.1	7.48
95.11	263804	1450	18926	18766	107.8	7.38
95.12	263988	1487	18731	17681	97.4	7.20
96.01	264162	1378	18927	16807	87.3	7.03
96.02	264295	1417	18639	18115	94.9	7.08
96.03	264485	1423	18676	18862	119.9	7.62
96.04	264701	1459	19265	19209	139.4	7.93
96.05	264878	1452	19652	20161	140.5	8.07
96.06	265072	1415	19653	21093	130.6	8.32
96.07	265284	1457	20269	19538	135.7	8.25
96.08	265498	1423	20053	21237	129.8	8.00
96.09	265723	1399	20195	21493	121.0	8.23
96.10	265943	1349	20057	21159	123.7	7.92
96.11	266130	1391	19903	19603	100.7	7.62
96.12	266313	1405	20264	18500	99.2	7.60
97.01	266490	1395	20471	18123	88.9	7.82
97.02	266620	1438	20507	19749	92.8	7.65
97.03	266789	1457	20421	20371	120.1	7.90
97.04	266993	1442	21342	21498	137.0	8.14
97.05	267177	1432	20613	21094	131.8	7.94
97.06	267368	1402	20896	22442	131.6	7.69
97.07	267575	1414	21574	21058	132.5	7.50
97.08	267792	1397	20873	21970	124.2	7.48
97.09	268014	1460	21213	22752	134.6	7.43
97.10	268234	1476	*	*	134.0	*

CASE 35

INTERRUPTING TINA

Tina is a graduate student adviser in the College of Business at the University of Michigan at Trenton. Her job is sometimes hectic, but she likes its flexibility. She sets her own schedule, which can vary from week to week if needed, as long as she gets the job done and works an average of 20 hours per week.

Tina is one of five advisers in the College of Business. Advisers have a myriad of job responsibilities that include counseling students, processing admissions, maintaining and updating student records, responding to requests from professors, and providing information to prospective students. Each adviser is trained to respond to all of these needs but also has a specific area of expertise. Tina is solely responsible for verifying student eligibility for graduation. She has to pull each student file to check whether all required courses have been completed and the student has a minimum grade point average of 3.0. This is a time-consuming process—about 200 students apply for graduation each semester.

Tina's office is located in the middle of the advising center, which is in the middle of the College of Business. She suspects that a quieter office in a corner somewhere would eliminate an awful lot of the noise, confusion, and extraneous requests. On the other hand, she needs to be available to answer quick questions for her colleagues and students. Tina is the senior student adviser with a great deal of knowledge. She is also an expert in the Student Information Computer System (SICS), which handles all of the student-related administrative record keeping. Often she's the only one on the floor who knows the ins and outs of the SICS.

Student advising is a big part of her job. Tina can schedule appointments in advance, or she might see students on a walk-in basis. It's tough to know from day to day how

These data were collected and provided to us by T. J. Gaub. The name of the institution has been changed.

many students will need to be served on a walk-in basis or how long their appointments will last. Nonetheless, the college has recently instituted a total quality management (TQM) program, which requires an eye toward customer (in this case, student) satisfaction. This means that students must now be seen on demand. Tina also advises students over the phone; advising sessions by phone can be either prescheduled or unscheduled.

In addition to these chores, Tina interacts with her supervisor, answers questions for professors, trains newly hired advisers, processes administrative paperwork, and answers phone calls.

Tina had more difficulty than usual last semester completing all of the graduation checks. It wasn't that there were more students graduating last semester. The problem seemed to be an unusually large number of interruptions. Even though the job calls for her do a variety of tasks beyond just graduation checks, things seem to be getting out of hand! Tina's supervisor is unable to pay any overtime, so the only solution for this semester is to identify and eliminate interruptions where possible.

To study this problem in more detail, Tina decided to perform an experiment. She created a worksheet on which she recorded every incident that took her away from graduation checks over a six-week period. The worksheet includes the time she clocked in and out, the type of incident, and the starting and ending time of the incident. The worksheet is shown below. At the end of the day, Tina calculated the length of each incident in minutes and categorized each incident by general type:

    A = SICS problems
    B = Work with coworkers
    C = Work with supervisor
    D = Answer phone (unscheduled)
    E = Clerical
    F = Student advising: walk-in
    G = Student advising: scheduled appointment
    H = Meetings
    I = Student advising by phone: scheduled
    M = Other

As the weeks went on, Tina's coworkers asked what she was doing with the clipboard and worksheet. Tina was mysteriously unresponsive to their questions.

Tina has asked you to analyze and interpret the data for her. Use whatever statistical technique you feel is appropriate to prepare a report.

WORKSHEET

DATE	TIME IN	TIME OUT	INTERRUPTION	START TIME	END TIME	LENGTH (mins)	TYPE
10/12	11:40	12:50	computer problem	11:52	11:53	1	A
			counsel Rick	11:55	11:56	1	B
			where is Betty?	12:09	12:10	1	C
			proof memo	12:13	12:14	1	C

(continues on the following page)

CASE 35 INTERRUPTING TINA

DATE	TIME IN	TIME OUT	INTERRUPTION	START TIME	END TIME	LENGTH (mins)	TYPE
			phone calls	12:20	12:23	3	D
			find file	12:23	12:50	27	E
10/13	3:15	4:55	check address	3:22	3:23	1	E
			phone calls	3:27	3:29	2	D
			give Rick message	3:29	3:29	1	B
			phone calls	3:50	3:52	2	D
			computer help	4:02	4:04	2	A
			phone call	4:26	4:26	1	D
			walk-in	4:33	4:35	2	F
			walk-in	4:36	4:40	4	F
10/14	1:45	5:00	question about answering machine	1:50	1:51	1	E
			question about file	1:52	1:54	2	E
			get memo for Betty	2:01	2:03	2	C
			walk-in	2:10	2:23	13	F
			SICS problem	2:25	2:26	1	A
			walk-in	2:59	3:15	16	F
			dean's signature	3:42	3:44	2	E
			walk-in	3:46	3:47	1	F
10/18	4:15	5:50	counsel Rick	4:16	4:20	4	B
			walk-in	4:27	4:28	1	F
			telephone	4:42	4:56	14	D
			walk-in	5:31	5:50	19	F
10/19	10:40	12:40	phone call	10:40	10:42	2	D
			phone appointment	10:47	11:02	15	I
			phone call	11:23	11:26	3	D
			phone call	11:38	11:39	1	D

(continues on the following page)

II–40   CASE 35   INTERRUPTING TINA

DATE	TIME IN	TIME OUT	INTERRUPTION	START TIME	END TIME	LENGTH (mins)	TYPE
			phone call	11:55	11:56	1	D
			phone call	11:56	11:58	2	D
			walk-in	12:05	12:14	9	F
			phone call	12:20	12:22	2	D
			walk-in	12:25	12:31	6	F
			phone call	12:34	12:35	1	D
			phone call	12:37	12:38	1	D
			phone call	12:38	12:40	2	D
10/27	2:30	5:00	counsel Rick	2:39	2:40	1	B
			forgot password	2:43	2:52	9	A
			appointment	2:55	3:05	10	G
			computer problem	3:05	3:21	16	A
			appointment	4:00	4:40	40	G
			appointment	4:45	5:00	15	G
11/2	10:30	12:45	SICS problem	10:31	10:38	7	A
			Mark steals the computer	10:38	10:41	3	B
			phone call	10:44	10:55	11	D
			check on professor	11:55	12:00	5	E
	2:10	6:00	walk-in	2:45	3:21	36	F
			phone appointment	3:33	3:40	7	I
			appointment	5:00	5:15	15	G
11/3	3:00	5:15	appointment	3:15	3:25	10	G
			appointment	4:15	4:30	15	G
			appointment	4:45	4:50	5	G
11/4	10:15	12:45	appointment	10:50	11:00	10	G
	2:00	5:00	walk-in	2:45	3:00	15	F
			walk-in	3:10	3:26	16	F

(concludes on the following page)

DATE	TIME IN	TIME OUT	INTERRUPTION	START TIME	END TIME	LENGTH (mins)	TYPE
			walk-in	3:58	4:05	7	F
11/8	4:00	6:30	appointment	4:05	4:30	25	G
			appointment	5:05	5:30	25	G
			training Shelly	5:35	5:40	5	B
			walk-in	5:50	5:56	6	F
11/9	9:45	12:30	appointment	9:50	10:20	30	G
			appointment	11:00	11:30	30	G
	2:00	6:30	personal call	3:10	3:20	10	M
			walk-in	3:50	3:55	5	F
			walk-in	3:55	4:10	15	F
			walk-in	4:30	4:35	5	F
			appointment	5:00	5:25	25	G
11/11	11:00	5:30	walk-in	1:15	2:30	75	F
			walk-in	3:06	3:08	2	F
			walk-in	4:24	4:26	2	F
11/12	3:00	5:30	Betty	4:00	4:15	15	C
			walk-in	4:45	4:50	5	F
11/15	4:00	6:30	appointment	4:00	4:15	15	G
			meeting	4:55	6:15	80	H
11/16	10:45	6:30	walk-in	11:15	11:16	1	F
			move furniture	12:30	12:45	15	M
			records	2:30	3:10	40	E
			phone call	4:40	4:55	15	D
11/18	10:15	12:15					
	1:15	5:15	walk-in	2:45	3:15	30	F
			walk-in	3:15	4:00	45	F
			walk-in	4:00	4:45	45	F

CASE 36

VIOLENCE ON TV—II

As the general manager of KTDS, the NBC affiliate in Tidusville, Oklahoma, Chris has a range of responsibilities that include programming, personnel, advertising, and public relations. His least favorite activity is responding to customer complaints. Unfortunately, there's been an unusually large number of complaints in the past few months from viewers and advertisers alike.

Most of the recent comments are objections to the level of violence on KTDS programs. Chris is sensitive to this issue because he has observed a gradual increase in violence on TV over the past 20 years. Chris really prefers the old-time movies in which dirty deeds were neatly sanitized and violent crimes occurred behind the scenes. He is sympathetic to the recent callers.

Nonetheless, he's in a tight spot. Chris knows that small, vocal groups do not necessarily represent the population at large. People who feel strongly about an issue are likely to speak out, while those who are content tend to remain silent. While the recent callers have denounced the level of violence on KTDS shows, Chris knows he must understand and serve *all* of the KTDS viewers.

Chris has a suspicion about the source of the recent calls. Four months ago, a flamboyant politician announced his candidacy for mayor. This candidate has received a great deal of air time on the local news due, in part, to his impassioned outbursts. Some people love him, others despise him, but almost everyone tunes in to the evening news in hopes of catching the latest controversy. One continuing theme of his platform is violence in America in general and violence on KTDS in particular. Over the past four months, the candidate has suggested that those opposed to violence in the media "let their voices be heard." Chris suspects that this fellow has inspired a large portion of the recent complaints to KTDS.

The survey design and data set compilation were undertaken by Ann Lee Bailey, an MBA student at the University of Colorado at Denver. The scenario has been altered to preserve the anonymity of the survey respondents.

Chris needs to sort all of this out. To understand the views of all KTDS patrons, he has commissioned an independent consulting firm to devise and administer a survey. The survey results are in, and he wants you to analyze the data (which reside in the file named VIOLENCE). Below is a copy of the questionnaire.

The survey was done over a three-week period in October. Two hundred phone calls were made and 106 people declined the invitation to participate in the survey. Of the 94 participating respondents, two people were offended by the question of income and refused to answer that particular question. Nonresponses are indicated in the data set by an asterisk. A random selection of phone numbers from the Tidusville phone book was used to select the sample.

What is Tidusville's perception of the level of violence on KTDS? Does one's perception of violence on TV vary with gender, age, marital status, income, or education? Do parents with children at home have a different tolerance for violence than those without children at home? Do viewers who spend a lot of time watching TV become desensitized to violence? Chris needs a report as soon as possible.

SURVEY

1. Gender   (0) male   (1) female

2. Age   (1) under 20   (2) 20-30   (3) 31-40   (4) 41-50   (5) over 50

3. Marital status   (0) married   (1) single or divorced

4. Do you have children at home?   (0) yes   (1) no

5. Household Income   (1) under $20,000   (2) $20,000-40,000   (3) $40,000-60,000   (4) over $60,000

6. Education   (1) high school   (2) some college   (3) college graduate   (4) graduate school

7. How many hours per week do you watch TV?
   (1) 0-7   (2) 8-14   (3) 15-21   (4) 22-28   (5) 29-35   (6) 36-42   (7) 43 or more

8. In your opinion, how violent are most TV programs?
   (1) much too violent   (2) somewhat too violent   (3) violent   (4) a little violent   (5) not very violent

DATA SET

GENDER	AGE	MARRIED	CHILDREN	INCOME	SCHOOL	HOURS	VIOLENT
0	3	0	1	2	4	2	5
1	2	1	1	1	2	2	3
1	4	0	1	3	1	3	1
1	3	0	1	3	3	4	2
1	4	0	0	4	4	1	2
1	5	0	0	*	4	1	1
1	3	1	0	2	3	1	1
0	3	0	0	3	2	3	5
1	3	0	0	2	4	5	1
1	4	0	1	4	4	5	4
1	5	0	1	2	1	2	3
1	4	0	0	4	2	1	1
1	3	1	1	2	2	1	3
1	3	1	0	2	3	2	3
1	2	0	1	3	3	4	4
1	4	0	1	2	2	2	2
1	2	1	1	2	4	1	4
1	3	0	1	2	4	2	2
1	2	1	1	1	2	3	3
0	3	1	0	2	2	3	3
1	2	1	1	1	2	3	2
0	4	1	0	2	3	4	2
1	3	1	1	2	3	1	4
1	3	1	0	2	2	3	2
1	3	0	0	3	3	4	1
1	2	1	1	1	3	1	1
0	2	0	1	2	4	1	1
0	2	1	1	2	4	2	2
1	2	0	1	4	3	1	2
1	2	0	1	2	4	2	3
1	2	1	1	2	4	1	4
1	2	0	0	3	3	1	1
1	4	1	0	2	2	3	1
1	3	1	0	3	4	3	1
1	4	0	0	3	4	1	1
0	2	1	1	1	2	4	5
1	2	1	1	2	2	1	3
0	4	1	1	2	3	1	2
1	5	1	1	1	2	3	3
1	4	1	0	2	3	1	2
1	2	1	1	2	2	1	1
0	3	1	1	2	3	2	3
1	5	1	1	2	4	1	1
1	4	0	1	3	2	5	1
1	3	1	1	2	3	4	3
1	3	1	0	2	3	4	3

(concludes on the next page)

CASE 36 VIOLENCE ON TV—II

GENDER	AGE	CHILDREN	MARRIED	SCHOOL INCOME	VIOLENT HOURS		
1	3	1	1	2	3	4	4
1	3	1	1	2	4	1	1
1	2	0	1	1	4	4	2
1	3	1	1	*	2	2	1
0	5	0	1	4	2	4	4
0	5	1	1	1	4	2	1
0	4	0	0	3	3	3	3
0	2	0	0	2	2	3	4
1	2	0	0	3	2	2	2
0	3	0	1	3	3	2	1
1	4	0	0	4	4	1	1
0	2	1	1	1	1	4	4
0	3	0	0	3	3	2	2
1	4	0	0	2	3	2	3
1	2	1	1	1	3	3	1
1	4	1	1	1	2	2	1
0	3	0	1	2	2	4	4
0	1	1	1	1	1	4	5
0	3	0	0	1	2	1	2
1	4	1	1	2	3	1	2
1	2	1	0	1	2	1	3
1	2	0	1	2	2	2	4
0	2	0	1	2	2	2	2
0	1	1	1	1	1	5	5
1	2	1	1	2	3	1	3
1	4	1	1	1	3	2	1
1	3	1	1	1	3	2	1
1	3	0	0	3	3	2	2
1	3	0	1	3	4	2	2
0	3	0	0	3	3	3	2
0	3	0	0	3	3	3	4
0	2	1	1	1	1	4	5
0	4	1	1	2	2	3	4
0	4	1	0	2	3	1	3
0	3	0	0	4	3	1	1
1	3	0	0	2	2	4	2
0	2	1	1	2	3	3	3
1	2	0	0	2	3	2	3
0	2	1	1	1	2	5	5
1	2	0	1	1	3	1	3
1	2	0	0	1	1	2	3
0	2	0	1	2	1	4	4
0	2	0	1	1	4	3	1
1	2	0	1	1	4	2	4
1	2	1	1	1	2	4	4
1	3	1	1	2	2	5	4
0	2	1	1	2	2	3	4
0	2	0	1	1	4	2	2

CASE 37

COMMERCIAL PROPERTY

In early 1992, the Arapahoe Assessor's office made local headlines. Arapahoe County, adjacent to Denver County in Colorado, had just undergone its annual property assessments. Many people thought the assessed values were much too low. If this were true, it would have implications for funding of schools, the perceived value of the properties and property tax revenues for the county. The Arapahoe County Assessor was in the center of all of this—he was responsible for the low assessed values and he owned one of the properties.

Is there any way to determine whether the assessments were fair? To understand how buildings might be assessed, you have collected information on recent selling prices of commercial property in Arapahoe County. Of course, the selling price need not equal the assessed value of the property, but in most jurisdictions the assessed value is supposed to be proportional to the fair market value of the property. Selling price is an obvious measure of that value.

Understanding the determinants of selling prices of commercial property is difficult. Private residences typically range in size from several hundred to several thousand square feet, but commercial property can be the corner drug store or the high-rise downtown. The expected *use* of the commercial property may influence its selling price; buildings designed to be restaurants, warehouses, or barbershops probably sell for different amounts. Like residential property, though, it might be expected that older commercial property sells for less, all else the same.

The data file, named COMMPROP, contains data on selling prices and physical characteristics of 75 commercial properties sold in Arapahoe County from July 1988 to June 1990. Do these data suggest which factors determine selling prices of

These data were compiled and analyzed by George Carr, a tax appraiser and consultant for the Colorado Division of Property taxation; Rochelle Hrigora Hass; Lisa E. Pieper, Research Associate; and Mark Truitt. The data come from "Arapahoe County 1992 Property Assessment Study Final Report," by Thimgan & Associates, Inc., conducted for the Colorado Legislative Council, Denver, Colorado, September 15, 1992, p. 110.

commercial property and, hence, which factors should be included in a fair assessment process? Is there seasonality to selling prices of commercial property? If so, how might this seasonality influence the ability to use selling price as a proxy for assessed value? Prepare a report of your findings.

DEFINITIONS

	CLASS	Primary use of the property: 0 = Merchandising 1 = Lodging 2 = Offices 3 = Special purpose 4 = Warehousing 5 = Multi-use 6 = Manufacturing
	LANDSIZE	Size of the land parcel in square feet
	TOWNSHIP	Location by township within Arapahoe County: 0 = Greenwood Village 1 = Aurora 2 = Englewood 3 = Littleton 4 = Unincorporated
	SQFEET	Number of square feet in the building
	FLOORS	Number of floors in the building
	DESIGN	Specific property characteristics impacting its use: 0 = Store/Retail 1 = Hotel 2 = General office 3 = Medical office 4 = Restaurant 5 = Garage/Auto 6 = Warehouse 7 = Industrial 8 = Other 9 = Commercial condominiums
	QUALCON	Quality of construction as assessed by an independent audit firm: 0 = Fair 1 = Average 2 = Good
	BUILT	Year the building was built (19XX)
	SYEAR	Year the building was sold (19XX)
	SMONTH	Month the building was sold (1 = January, etc.)
	PRICE	Selling price of the property in dollars

DATA SET

CLASS	LANDSIZE	TOWNSHIP	SQFEET	FLOORS	DESIGN	QUALCON	BUILT	SYEAR	SMONTH	PRICE
6	24393	2	11000	1	7	0	76	88	12	295000
3	8235	2	2400	1	7	0	75	89	5	120000
6	35912	2	23668	1	7	0	73	89	3	395000
6	28000	2	11567	1	7	0	70	89	6	280000
4	18229	2	6885	1	6	0	79	90	2	195000
6	14840	2	10000	1	7	0	71	90	2	205000
6	13160	2	8550	1	7	0	74	90	4	201000
6	43065	2	10240	1	7	0	73	88	12	350000
4	21587	2	11280	1	6	0	78	89	10	300000
4	45000	2	25224	1	6	0	78	89	2	425000
4	17480	2	6900	1	6	0	85	89	8	207500
2	6250	2	3300	1	2	0	80	89	10	230000
1	48260	1	43228	3	1	1	72	89	11	600000
0	15000	1	7450	1	0	0	42	90	3	250000
3	25000	1	3864	1	4	1	70	88	11	200000
3	41136	1	3478	1	5	1	73	89	7	350000
3	37635	1	3544	1	4	1	60	90	2	217000
5	20000	1	5640	1	0	1	78	89	7	262500
0	13983	1	2400	1	0	1	77	90	1	241250
2	79933	1	38924	3	2	1	82	89	12	1300000
1	267737	1	187528	7	1	2	82	89	3	13700000
0	17550	1	5040	1	0	0	76	89	12	182250
0	9828	4	2490	2	2	0	65	89	2	80000
2	49648	1	8776	1	2	0	77	89	3	400000
3	60364	1	8641	2	4	1	79	89	6	618000
2	1024	1	1390	2	9	1	80	90	4	92500
2	1024	1	1408	2	9	1	80	90	1	85000
2	1152	1	1440	2	9	1	84	88	12	99500
2	1152	1	1440	2	9	1	84	89	11	87000
2	1152	1	1408	2	9	1	83	90	4	83000
2	1152	1	1408	2	9	1	83	90	5	84000
3	233438	1	44562	1	0	0	86	89	8	2100000
2	11644	1	3812	1	9	1	84	90	6	102500
2	25655	1	9352	2	2	1	80	89	3	759000
2	54320	1	32130	3	2	1	83	89	4	1450000
4	30381	1	12980	2	6	0	81	88	7	300000
2	5502	1	1625	1	9	1	82	89	5	58100
4	28070	1	8918	1	6	1	85	89	12	175000
4	102148	1	11603	1	6	1	86	90	6	285000
5	107157	1	34122	1	0	0	79	90	4	850000
0	106252	1	35836	1	0	1	78	89	12	1075000
2	4800	1	4170	1	2	1	81	89	11	165000
2	6300	1	5571	1	2	1	81	89	9	350000
4	44012	4	12843	2	6	1	82	89	8	400000

(concludes on the next page)

CASE 37 COMMERCIAL PROPERTY

CLASS	LANDSIZE	TOWNSHIP	SQFEET	FLOORS	DESIGN	QUALCON	BUILT	SYEAR	SMONTH	PRICE
4	41528	4	8400	1	6	0	84	89	1	232000
5	23460	0	5480	1	0	1	71	89	8	310000
2	104548	0	59854	4	2	1	74	88	8	3350000
2	87991	0	14764	1	2	1	65	90	4	810000
2	85410	0	10492	1	2	1	68	88	10	600000
2	193321	0	77792	3	2	1	80	89	7	3325900
3	174240	3	48887	3	3	2	85	89	1	1440000
2	191403	4	111231	3	2	1	82	89	12	4683300
2	87120	4	19174	2	2	1	85	89	9	652350
4	266108	4	81190	1	6	1	82	90	5	2100000
4	6400	4	6292	1	2	1	80	88	11	227000
4	8000	4	7880	1	8	1	79	89	8	300000
4	4800	4	4704	1	2	1	79	90	2	189900
4	8000	4	7892	1	2	0	79	89	11	200000
3	20034	2	8352	2	3	1	65	88	9	475000
3	81295	2	39218	6	3	1	78	89	1	700000
5	15625	2	6084	1	0	1	85	89	9	513000
0	4625	2	1921	2	0	0	30	89	7	85000
0	25000	2	6080	1	0	0	60	89	2	260000
4	24000	2	10260	1	6	0	83	88	7	550000
3	21978	2	6080	1	5	1	88	89	8	600000
0	28828	3	910	1	0	0	65	90	3	90000
3	24700	3	5365	1	0	0	75	89	2	300000
0	5625	2	4500	1	0	0	53	89	1	105000
3	84040	3	31461	3	3	1	66	89	11	650000
0	4000	3	4489	1	0	0	60	89	4	105000
3	16800	3	3478	1	3	1	71	88	8	215000
5	267300	3	89820	1	0	1	56	90	6	1900000
4	217800	3	74550	1	6	1	85	90	5	1900000
0	24137	3	9775	1	0	1	81	88	9	440000
5	91500	3	18850	1	0	1	78	89	11	640000

CASE **38**

AIDS RATES

Our intuitive impressions of rates are often erratic or wrong, particularly about issues that are sensitive, emotional, or scary. For example, ask 10 or 20 people to estimate the rate of acquired immunodeficiency syndrome (AIDS) cases in the United States, first for males, and then for females, and record their answers in units of, say, cases per 1000 people. Prepare an appropriate summary of their responses. Record your own estimates, too.

Now use the World Wide Web to find the latest data from, say, the Centers for Disease Control and Prevention, about AIDS rates in the United States, and prepare an appropriate summary of *that* data. How accurate were the opinions of the people in your sample? How consistent?

CASE 39

SOFTWARE SERVICE CALLS

"What can you do?" asked Sam. "The number of service calls we get in any week or month jumps around so much, it's no wonder that sometimes we seem to be overstaffed, while at other times we are completely swamped and can't get to all the reported problems anywhere near as quickly as we ought to." Sam knew his supervisor wouldn't like his explanation much, but he still felt it was an accurate description of the problem he faced.

Sam managed a department in the national service center for a software development company. As part of the customer service and warranty operation, the national service center staffed a hotline, to which customers could report discrepancies or problems. Customers could file formal, written trouble reports (STR), but Sam was concerned about telephoned trouble reports, not STRs.

His department received direct service requests by phone (SCALLs) for one specific product in the firm's catalog, a fairly large package for computer-aided design and drafting. The package had been sold for 18 months. His problem was to have the appropriate number of service representatives on hand to provide decent response time to customers. "Decent" was not defined in any specific terms (at least not yet), but Sam agreed with his supervisor that performance lately had certainly been unsatisfactory, at least in spurts. Sam knew that the allocation of service representatives to different products could in principle be changed from time to time. The difficulty was how to know in advance how many would be needed for which product.

Sam stared at the numbers on his desk and remembered the tone in his supervisor's voice. He had to figure out something. The numbers on his desk are also available in

---

This case describes a real business situation. The data are used here with the permission of the students who originally provided them to us.

file SERVCALL. Q is the quantity of the product sold in each month, and MONTH is a count of calendar months elapsed since the product's introduction. The other variables are defined above.

DATA SET

MONTH	Q	STR	SCALL
1	20	28	76
2	19	18	63
3	28	10	79
4	37	53	73
5	2	47	89
6	91	41	97
7	4	38	68
8	11	34	118
9	45	66	85
10	4	28	88
11	9	63	99
12	14	58	68
13	10	40	70
14	8	43	84
15	7	41	85
16	9	38	72
17	11	32	74
18	8	33	72

CASE **40**

TITANIC SHIPPING COMPANY

Titanic Shipping Company depends on its nationwide system of computer terminals for entering orders, tracking shipments, and other routine functions. Some 1500 terminals and personal computers in more than 70 locations are connected through a state-of-the-art telecommunications network to central computing facilities in Newark, New Jersey. Thanks to good backup procedures and steady improvements in communications technology, the system has become reliable, and failures (in the strict sense of the term) have been rare.

The performance of the system has troubled the users, though. At any point during a normal day, several hundred users may be entering or retrieving information about the company's shipments or updating customer records. When response time for these transactions deteriorates, users can't work effectively. Recently, they've been letting the computer center manager know their feelings—forcefully.

The computer center manager understands their concerns, but isn't sure how to respond. To some extent, of course, improving response time requires things like new facilities or better performance tuning of their systems, both of which take money. That can be negotiated in the normal company budget process. Right now, he is more concerned about losing his credibility with the users he is supposed to serve. Several years ago, the users' representatives agreed that response times of two seconds would be adequate. Since then, the computer center has tracked average response time, as measured by an internal tool in the computer's operating system. According to this measure, performance has been generally satisfactory, at least for large portions of the time. Still, the users complain that the system is not meeting their needs. The manager realizes that part of the problem is one of "translation": the users don't know how to specify exactly

---

The company name and the specific data for this case are simulated, but the situation corresponds closely to a number of real situations with which we are familiar.

how the computer center ought to measure response time (at least not in any precise, quantifiable terms), but when performance is bad, they know it, and they complain. As long as the computer center continues to track performance by using a measure that (apparently) is unrelated to the users' perceptions of their needs, relations between the users and computer center management will probably continue to deteriorate.

During the last quarter the computer center has developed a system that asks each user to rate the system performance as "OK" or "Bad" before signing off. Thirteen weeks' worth of data from the 10:00–11:00 hour for each workday are given below and in file TITANIC. The 10:00 hour is usually a busy one, and generates many of the user complaints. Data from two holidays have been omitted.

The manager wonders if these data can provide some insight that will allow him to develop a better measure of response time. He really wants to know two things:

- How should he measure response time?
- What target or standard performance should he use?

The measurements in the data set come from the software in the operating system and could be obtained routinely if it were helpful to do so. In setting the standard, the manager wants to use a value that corresponds as closely as possible to the users' perceptions, so that when the computer center's own measurements indicate acceptable performance, the users will agree. Can you advise him?

DEFINITIONS

AVERESP	Average response time during that hour (seconds)
RESP50	50th percentile of the response times during that hour (seconds)
RESP75	75th percentile of the response times during that hour (seconds)
RESP90	90th percentile of the response times during that hour (seconds)
RESP95	95th percentile of the response times during that hour (seconds)
NBAD	Number of users who rated performance "bad" when signing off
NOK	Number of users who rated performance "OK" when signing off

CASE 40 TITANIC SHIPPING COMPANY

DATA SET

DAY	AVERESP	RESP50	RESP75	RESP90	RESP95	NBAD	NOK
1	2.223	1.559	2.982	4.944	6.426	322	5
2	1.253	0.989	1.825	3.008	3.951	2	276
3	1.758	1.157	2.257	3.734	4.882	144	155
4	1.703	1.287	2.632	4.238	5.593	310	4
5	1.395	1.119	2.174	3.682	4.764	104	191
6	2.380	1.928	3.757	6.228	8.017	343	7
7	2.486	1.694	3.384	5.695	7.366	338	1
8	2.049	1.556	3.039	5.034	6.523	326	3
9	2.771	1.903	3.822	6.293	8.147	351	1
10	2.132	1.734	3.367	5.482	7.182	337	2
11	2.566	1.785	3.621	5.948	7.633	343	3
12	2.090	1.488	2.888	4.856	6.277	314	10
13	3.032	1.819	3.686	6.044	7.879	346	2
14	1.029	0.776	1.556	2.504	3.270	15	247
15	1.465	1.080	2.028	3.342	4.314	5	283
16	2.144	1.558	3.083	5.090	6.612	317	13
17	1.211	0.951	1.928	3.211	4.168	2	281
18	0.963	0.797	1.618	2.639	3.364	5	261
19	2.800	2.011	4.043	6.736	8.708	356	1
20	2.507	1.629	3.327	5.430	7.027	315	23
21	2.387	1.661	3.327	5.479	7.132	328	10
22	3.118	2.085	4.078	6.697	8.696	357	1
23	2.182	1.851	3.590	5.981	7.743	310	35
24	1.812	1.807	3.556	5.934	7.699	343	1
25	0.916	0.861	1.603	2.603	3.347	7	258
26	2.754	2.053	4.188	6.813	8.907	355	6
27	2.592	1.768	3.533	5.758	7.571	341	3
28	3.135	2.085	4.061	6.755	8.692	340	18
29	3.076	2.121	4.213	6.868	8.974	353	8
30	1.795	1.207	2.421	3.960	5.179	263	43
31	2.795	2.089	4.090	6.813	8.807	355	3
32	1.803	1.186	2.211	3.710	4.811	128	169
33	1.786	1.374	2.662	4.283	5.562	312	3
34	2.733	2.090	4.160	6.935	9.043	352	8
35	2.880	2.086	4.037	6.662	8.663	353	4
36	1.589	1.081	2.052	3.432	4.444	31	258
37	1.682	1.346	2.580	4.192	5.436	306	6
38	0.902	0.785	1.551	2.493	3.269	4	257
39	1.045	0.898	1.701	2.841	3.657	4	267
40	1.194	1.009	1.963	3.208	4.091	20	265
41	2.306	1.696	3.291	5.421	7.020	334	3
42	1.329	1.111	2.125	3.474	4.432	7	286
43	3.016	1.937	3.724	6.147	8.060	348	1
44	2.477	1.668	3.377	5.547	7.185	330	9
45	1.567	1.187	2.282	3.797	4.916	185	115

(concludes on the next page)

CASE 40 TITANIC SHIPPING COMPANY

DAY	AVERESP	RESP50	RESP75	RESP90	RESP95	NBAD	NOK
46	1.617	1.183	2.355	3.864	5.005	221	82
47	2.440	1.713	3.455	5.699	7.405	339	2
48	1.741	1.402	2.821	4.574	6.007	315	6
49	2.028	1.369	2.730	4.392	5.776	317	1
50	1.776	1.464	3.008	4.956	6.401	303	25
51	1.116	0.831	1.633	2.675	3.508	2	265
52	2.912	1.924	3.857	6.409	8.249	351	1
53	1.276	0.898	1.768	2.904	3.756	3	271
54	1.392	1.312	2.523	4.186	5.416	303	7
55	2.142	1.863	3.767	6.175	7.997	336	14
56	1.533	1.156	2.148	3.590	4.648	64	230
57	2.311	1.704	3.486	5.714	7.395	335	7
58	2.464	1.851	3.713	6.054	7.867	329	20
59	2.551	2.126	4.186	6.887	8.882	358	3
60	1.396	0.853	1.536	2.516	3.355	1	259
61	2.876	2.030	4.060	6.628	8.601	350	8
62	2.451	1.723	3.273	5.468	7.102	318	18
63	1.924	1.446	2.743	4.625	5.942	316	2

CASE 41

CAMPAIGN STATISTICS

"That's good!" said the congressman. "We can use that to show the voters just what happens when our opponents stay in office for four years." He was referring to data that showed that the effective income tax rate had increased from 14.1% to 15.2% of adjusted gross income during the period 1974 to 1978.

"What I'd like, though," he went on, "is to break it down into smaller categories, like, say, $10,000 to $14,999. That way, we can kill two birds with one stone. First, we can choose the category that looks the worst for our opponents. Second, the people in that category will know that we're talking about them, not just some 'average' or someone a lot richer or poorer than they are. See what you can work up for me!"

With that, he was off to a campaign appearance. The data he referred to are given below. Provide the breakdown he wants.

TABLE

TOTAL INCOME AND TOTAL TAX (IN THOUSANDS OF DOLLARS)

Adjusted gross income	1974 income	1974 tax	1978 income	1978 tax
Under $5,000	41,651,643	2,244,467	19,879,622	689,318
$5,000–$9,999	146,400,740	13,646,348	122,853,315	8,819,461
$10,000–$14,999	192,688,922	21,449,597	171,858,024	17,155,758
$15,000–$99,999	470,010,790	75,038,230	865,037,814	137,860,951
$100,000 or more	29,427,152	11,311,672	62,806,159	24,051,698
Total	880,179,247	123,690,314	1,242,434,934	188,577,186

This case is based on real data, originally reported by C. H. Wagner in *The American Statistician*, Volume 36, No. 1, February 1982, pp. 46–48. Reprinted with permission from *The American Statistician*. Copyright 1982 by the American Statistical Association. All rights reserved.

The effective tax rate for a given category is the tax paid divided by the total income reported in that category. For example, for 1974 the effective tax rate in total was 123,690,314/880,179,247=.141 and in 1978 the corresponding number was 188,577,186/1,242,434,934=.152. These are the numbers to which the congressman was referring before he left for the campaign trail.

CASE **42**

HOSPITAL CHARGES

When administrative duty's to be done, the hospital manager's lot is not a happy one. For example:

• Physicians have control over a substantial fraction of hospital costs, including the length of time a patient spends in the hospital and which drugs and other supplies are used in the patient's treatment.

• On the other hand, the physicians are usually *not* employees of the hospital, so while the hospital is responsible for managing the costs, its administration has rather less control over those costs than it does in many other business settings. Hospitals *do* generally grant "admitting privileges" to individual physicians and review their performance on a periodic basis, but short of rescinding those privileges (or threatening to do so), the hospital has little direct influence over many expenditures.

• Further, the hospital's revenues are determined largely by the patient's insurance coverage, which may pay either a more or less fixed amount based upon the "Diagnosis Related Group" (DRG, a classification system used by many federal and other programs), or a percentage of the bill (in more "classical" kinds of insurance schemes). To what extent differences in insurance coverage may or may not influence the patient's treatment or quality of care is a matter of occasional controversy.

Being sure that analyses of physician performance data are fair and balanced is important in any such performance review. The 289 patient records in the data for this case (available in file HOSPITAL) are for normal deliveries of babies. DAYS is the number of days the patient spent in the hospital, CHRGS is the total expenses charged to that patient, PHYS is a code identifying the physician, and PAYOR indicates the type of insurance the patient carried: 1 for managed care insurance and 0 for commercial insur-

These data are real, but the name of the hospital and of the person who collected them are confidential.

CASE 42 HOSPITAL CHARGES

ance. Other things being equal, do there appear to be any substantial differences in charges among the various physicians? Write a brief report summarizing your findings.

DATA SET

DAYS	CHRGS	PHYS	PAYOR	DAYS	CHRGS	PHYS	PAYOR
2	1701	10	0	2	929	13	1
2	3500	4	1	3	3464	14	1
2	3165	10	0	2	3173	10	0
1	1953	14	1	2	3034	14	1
2	2607	2	1	2	3716	12	1
2	2944	12	1	2	2137	6	1
4	5063	2	1	3	3430	6	1
2	2396	13	1	3	3041	6	1
4	4903	2	1	2	2247	12	1
2	2280	7	1	3	4357	12	1
2	2939	7	1	2	3050	12	1
3	3585	4	1	2	2779	12	1
2	3073	13	1	2	2357	4	1
2	2880	14	1	2	2620	7	1
2	2358	10	0	3	3369	13	1
3	3418	2	1	2	3090	7	1
3	3604	2	0	4	6340	11	1
3	3480	6	1	2	2900	10	1
2	3306	10	0	3	3709	2	1
3	3047	4	1	3	3503	4	1
2	2157	11	1	2	2991	12	1
2	2439	4	1	2	2941	14	1
3	2647	11	1	2	2668	14	1
2	1874	6	1	2	2138	2	1
1	2139	10	1	2	2227	12	0
2	2424	14	1	2	2838	10	0
2	2450	12	1	2	2681	2	0
2	2198	14	0	3	3945	7	1
2	3355	10	1	2	2026	4	0
2	2609	4	0	3	3160	10	1
2	2324	2	1	2	3270	12	0
2	2207	12	1	3	4146	6	1
2	2649	11	0	2	3757	11	1
3	3375	10	1	2	2435	7	1
2	3906	6	1	2	2343	10	1
3	4049	14	1	3	3932	2	1
2	2564	13	1	2	2285	10	1
2	2745	11	1	2	3001	14	1
3	1254	6	1	2	1864	7	0
2	2953	2	1	3	4711	12	1
2	2334	7	1	3	3283	2	0
2	2809	7	1	2	2592	7	1

(continues on the following page)

CASE 42 HOSPITAL CHARGES

DAYS	CHRGS	PHYS	PAYOR	DAYS	CHRGS	PHYS	PAYOR
1	2081	12	1	2	2804	2	0
1	1791	14	1	2	2801	13	0
2	3729	2	1	2	2638	7	1
2	2117	10	1	2	2263	10	0
3	4510	11	1	2	2449	11	0
2	2017	7	1	2	2473	11	0
2	3118	14	1	3	2864	6	1
3	3392	2	1	2	2534	11	1
2	1854	4	1	2	2550	10	1
14	14898	2	1	2	3768	12	1
3	3819	2	1	2	3194	10	0
3	2644	4	1	1	2153	12	0
3	2331	13	0	1	1840	7	0
3	3492	11	1	2	2522	12	1
2	2666	7	0	1	2145	12	1
3	4248	2	1	2	2403	6	1
3	2500	4	1	3	3287	2	1
2	3112	10	1	2	2979	6	1
3	2955	7	1	2	3182	7	0
2	2761	14	1	2	2954	10	0
2	1905	2	1	2	2158	11	0
3	3394	10	1	2	3455	12	0
2	3663	10	1	2	2367	14	1
3	3059	6	1	3	2798	10	1
2	2528	10	1	2	2408	11	1
2	2823	2	1	2	2468	11	1
6	6710	11	1	3	3081	12	1
2	2543	13	1	2	2218	7	1
3	2785	2	0	3	2133	11	0
2	2589	12	0	4	5499	13	1
2	3183	11	1	2	2251	12	1
1	1806	10	1	2	2310	11	0
2	3115	10	0	1	2220	11	0
2	3204	7	1	2	2176	10	0
2	2921	2	0	1	2048	2	1
3	4933	2	0	2	2566	11	1
2	4722	13	1	2	2450	11	0
3	5633	13	1	2	2582	11	1
2	2780	10	0	2	2612	7	0
2	2798	4	1	2	2683	6	1
3	2066	7	1	2	2827	14	0
3	2494	11	1	2	3202	10	1
3	2724	10	0	3	3779	10	1
2	2352	11	0	2	2833	12	1
1	1793	7	1	2	3458	10	0
2	2629	4	1	1	2789	7	1
2	3094	14	0	2	1590	7	1

(continues on the following page)

CASE 42 HOSPITAL CHARGES

DAYS	CHRGS	PHYS	PAYOR	DAYS	CHRGS	PHYS	PAYOR
3	3868	7	1	2	3032	7	0
2	3077	12	1	3	2891	4	0
2	3109	10	0	1	2886	10	1
3	2697	4	0	1	3381	2	0
2	3182	13	1	2	2752	10	0
2	3034	6	1	1	3675	13	1
1	2179	14	1	2	2840	13	1
1	2178	7	1	2	3105	13	0
3	4185	11	1	2	2288	11	1
2	1567	10	1	2	3128	14	0
2	2347	14	1	2	2289	7	0
1	2268	10	1	2	2602	10	0
3	3617	2	0	2	2604	14	1
2	2858	10	1	2	1023	12	0
2	2542	10	1	1	1899	10	0
2	2665	7	1	2	2334	10	1
2	2436	6	1	2	2310	2	1
2	2392	12	1	3	3636	6	1
2	2739	7	0	2	4206	7	0
3	2181	12	0	2	2414	12	1
2	2308	4	0	2	2206	13	1
2	3800	13	0	3	2907	2	1
2	2242	12	1	2	3017	14	1
2	2314	7	1	2	3050	14	1
1	1960	6	0	2	1973	7	1
3	4183	11	1	2	2692	10	1
1	1663	4	1	3	2564	11	0
1	2070	7	0	2	3693	10	1
3	2753	6	1	2	2741	13	1
3	2209	6	0	2	2888	2	1
3	3218	10	1	2	3279	6	0
2	2419	13	0	3	2873	7	1
3	3289	11	0	2	2582	10	1
2	2222	4	1	2	2640	2	0
1	2741	4	0	1	2068	14	1
3	3454	10	1	2	2123	10	0
2	3848	10	1	3	3480	14	1
2	2292	7	0	3	4475	6	0
2	2594	7	1	2	2312	14	1
2	2378	6	1	2	2898	4	1
1	902	7	0	3	3708	10	0
3	2935	7	1	1	2078	10	0
2	3230	6	1	2	2570	11	1
2	2219	2	0	3	3639	14	1
3	2687	12	1	2	4674	12	0
2	1848	14	1	2	2534	14	0
3	2891	4	0	2	2797	7	0
2	2698	12	0	2	3139	12	0

(concludes on the following page)

DAYS	CHRGS	PHYS	PAYOR
3	3826	2	0
2	2067	10	0
2	2840	2	1
2	2465	12	1
1	1947	12	0
1	2167	7	0
2	2840	10	0
2	2711	6	1
2	3137	2	0
1	2955	2	0
2	2563	10	1
1	1924	4	1
1	2062	6	0
2	2663	13	0
1	2184	2	0
2	3064	10	1

CASE 43

COAL TESTING

Coal is a heterogeneous bulk commodity, typically sold under contracted terms that include bonuses and penalties associated with various quality parameters and impurity levels of the coal. In many cases, predicting and controlling the quality of the coal accurately can mean the difference between economic success and failure for a coal mine.

The heating value is a quality of particular interest. It is the basis of payment in most coal supply contracts. In the United States, the heating value of coal is measured in British thermal units per pound (BTU/lb), by using a national standard test procedure (ASTM D-2015). ASTM standards also indicate that repeated measurements should give values within 100 BTU/lb.

Normally, the only way to determine heating value is to send a carefully prepared sample off to a laboratory and wait for the results. This can lead to disputes, if the customer and the supplier don't agree on which lab to use or if the results from different laboratories give drastically different results.

In an effort to determine the extent of any intra- or interlaboratory variation in heating value measurements, one coal company prepared 50 samples of coal from a common area (so that results were expected to be very close). They shipped the samples in two batches to five reputable laboratories. Each batch consisted of five samples sent to each lab; the two batches were sent at different times. The samples were drawn randomly, packed in water (to minimize oxidation effects) and sent in sealed plastic containers. The labs were not aware that an experiment was being run, and the coal company hoped that the results would thus reflect the labs' normal operating procedures.

This case is based on a real situation, described to us by H. Erik Scherer, P. E., Mining Engineer and Consultant. He also provided the data.

CASE 43 COAL TESTING

The labs made several measurements, but only those for heating value are given below. What do you think the coal company should conclude about the performance of the laboratories on its list, at least as far as heating value measurements are concerned?

The data are available in file COALLAB. The variables are: LABNO, the laboratory number; BATCH, the batch number (0 or 1); and BTU, the BTUs per pound as reported by the laboratory.

DATA SET

OBS	LABNO	BATCH	BTU
1	1	0	11655
2	1	0	11690
3	1	0	11654
4	1	0	11653
5	1	0	11665
6	1	1	11754
7	1	1	11676
8	1	1	11712
9	1	1	11711
10	1	1	11684
11	2	0	11665
12	2	0	11618
13	2	0	11572
14	2	0	11686
15	2	0	11725
16	2	1	11602
17	2	1	11582
18	2	1	11764
19	2	1	11618
20	2	1	11637
21	3	0	11586
22	3	0	11588
23	3	0	11599
24	3	0	11596
25	3	0	11622
26	3	1	11428
27	3	1	11429
28	3	1	11488
29	3	1	11632
30	3	1	11478
31	4	0	11651
32	4	0	11780
33	4	0	11663
34	4	0	11769
35	4	0	11800
36	4	1	11630
37	4	1	11587

(concludes on the following page)

OBS	LABNO	BATCH	BTU
38	4	1	11567
39	4	1	11531
40	4	1	11630
41	5	0	11591
42	5	0	11605
43	5	0	11549
44	5	0	11590
45	5	0	11609
46	5	1	11739
47	5	1	11747
48	5	1	11768
49	5	1	11768
50	5	1	11794

CASE 44

PERFORMANCE EVALUATIONS

Ann, an independent contractor who specializes in human resources issues, has been asked by the XYZ company to undertake an independent study of its employee performance evaluation practices. XYZ, a nonprofit organization, is a professional review organization in the health care industry. Most of its revenue comes from government contracts for Medicare and Medicaid utilization reviews.

The employee performance appraisal involves a fairly complex mix of subjective and objective scoring. Each evaluator fills out a form that covers 11 categories of employee performance, such as adaptability, accountability, communication and decision-making skills, sensitivity to cost-effectiveness issues within the organization, attendance, and training and development of other employees (see Exhibit A). Nonsupervisory personnel are evaluated on 8 of the 11 categories; supervisory and management employees are rated on all 11 categories. The same form is used for all employees, although the weight applied to each category varies with the employee's grade and responsibilities, as shown in Exhibit B. Weights are assigned in such a way that any employee can receive a final score anywhere from 0 to 200. Higher scores mean better performance.

Each employee has an evaluation initiated on the 12-month anniversary of his or her hire date, so evaluations go on at XYZ throughout the year. Any employee who is a new hire, has been transferred between departments, has been promoted, or has been demoted is considered to be on probation. These employees are given three-month evaluations in addition to the annual evaluations.

These data are real, although, because they deal with personnel issues, the name of the company has been disguised. Claudia Roberts Salvano collected the data set and provided it to us. Claudia received an M. S. degree in Management in 1993 and was the Manager of Human Resources at the time that she collected the data.

EXHIBIT A   EMPLOYEE PERFORMANCE APPRAISAL FORM

ANNUAL PERFORMANCE EVALUATION

Date Evaluation Due:_____   Time on Job:_____   Employment Date:_____
Employee: _____   Title: _____
Evaluated by:_____   Title: _____
Reviewed by: _____   Title: _____

RATING KEY

RATING	POOR(P)		FAIR(F)		COMPETENT(CP)		COMMENDABLE(C)		DISTINGUISHED	
INDIVIDUAL RATING	P–	P+	F–	F+	CP–	CP+	C–	C+	D–	D+
CATEGORY POINTS	1	2	3	4	5	6	7	8	9	10
TOTAL POINTS	0-50		51-70	71-90	91-110	111-130	131-150	151-170	171-190	191-200
Raise %	0		0	1%	2%	3%	4%	5 ¼%	5 ¾%	6 ½%

DESCRIPTION OF CATEGORY

Poor	Performance does not meet requirement of position.
Fair	Performance adequate most of the time. Marginal performer in one or more areas.
Competent	Responsibilities are handled competently. Meets all requirements of position.
Commendable	Knowledge and mastery of position performance marked by high quality.
Distinguished	Performance superior in all aspects. Quality leaves nothing to be desired.

Employee   Manager   Approved by
Date: _____   Date: _____   Date:_____

Employee Agrees _____   Employee Does Not Agree_____

EXHIBIT A *(continued)*

Annual _____
Three-month _____

EMPLOYEE EVALUATION

Employee Name _____   Due Date _____

1. ( )   **INITIATIVE/ADAPTABILITY/ACCOUNTABILITY**   Rating____ Pts____

    Demonstrates ability to work with others in organization. Adjusts to changed work assignments or conditions, accepts guidance and special assignments. Demonstrates capabilities for establishing work priorities and begins tasks on own. Fulfills the full range of accountabilities indicated in his/her job description.

    COMMENTS: _____
    _____
    _____

2. ( )   **COMMUNICATION**   Rating____ Pts____

    Clearly, concisely, and persuasively expresses ideas and instructions verbally and in writing and understands the ideas and instructions of others.

    COMMENTS: _____
    _____
    _____

3. ( )   **DECISION MAKING**   Rating____ Pts____

    Demonstrates the ability to evaluate and make proper decisions within a reasonable time frame. Evaluates data, reaches sound conclusions, makes appropriate recommendations, and follows through with timely decisions.

    COMMENTS: _____
    _____
    _____

4. ( )   **COST EFFECTIVENESS**   Rating____ Pts____

    Demonstrates and promotes an awareness of the importance of cost effectiveness and reduces expenses (through unnecessary labor and time, wasted supplies, unneeded equipment, etc.) and reduces costs when and where practicable.

    COMMENTS: _____
    _____
    _____

5. ( )   **QUANTITY OF TASKS**   Rating____ Pts____

    Performs day to day assigned activities and special projects, within prescribed standards and in a timely fashion.

    COMMENTS: _____
    _____
    _____
    _____
    _____

EXHIBIT A   *(continued)*

6. ( ) QUALITY OF TASKS   Rating____ Pts____

   Produces high quality work with minimal error rate.

   COMMENTS: _____
   _____
   _____
   _____
   _____

7. ( ) ADHERENCE TO CORPORATE POLICY AND DECISIONS   Rating____Pts____

   Understands the rationale used in decisions, accepts the final decision, and is able to properly convey the decisions to others when appropriate. Adheres to all Corporate Policies outlined in the Personnel Manual and related policy memos.

   COMMENTS: _____
   _____
   _____
   _____

8. ( ) ATTENDANCE   Rating____ Pts____

   Arrives at work on time, returns from breaks and lunch on time, schedules vacation time in advance, schedules Doctor and Dentist appointments so as to use the least amount of work time. The number of hours of sick leave used is a major component for this category.

   COMMENTS: _____
   _____
   _____
   _____
   _____

9. ( ) PLANNING AND ORGANIZING   Rating____ Pts____

   Effectively schedules and coordinates the flow of work. Organizes and utilizes his/her people, materials, and equipment. Organizes and utilizes his/her own time, meets established deadlines. Demonstrates ability to prepare annual budgets and to operate within approved budgets.

   COMMENTS: _____
   _____
   _____
   _____

10. ( ) MANAGING, CONTROLLING, AND DELEGATING   Rating____ Pts____

    Leads and motivates people to effective group and individual effort. Directs, reviews, and checks work of subordinates. Establishes and maintains standards of performance. Takes corrective action when necessary. Defines and delegates authority to others.

    COMMENTS: _____
    _____
    _____
    _____

EXHIBIT A *(concluded)*

11. ( ) DEVELOPMENT OF OTHERS   Rating___ Pts___

Encourages and assists individuals, through training and other methods, to acquire knowledge and skills essential for effective job performance. Selects individuals with potential for promotion and assists them to qualify for advancement with assurance that equal employment opportunity corporate goals are met.

COMMENTS: _____
_____
_____
_____

EXHIBIT B   PERFORMANCE EVALUATION CATEGORY WEIGHTS

Category	Director	Manager	RN	Technical	Secretarial	Clerical	Medical Director
Initiative/Adaptability/Accountability	3	3	3	3	2	2	3
Communication	2	2	2	2	2	2	2
Decision Making	2	2	3	2	2	2	2
Cost Effectiveness	2	1	2	1	1	1	1
Quantity of Tasks	1	1	3	4	4	4	2
Quality of Tasks	1	1	3	4	4	4	2
Adherence to Corp. Policy	2	2	2	2	2	2	2
Attendance	1	1	2	2	3	3	0
SUPERVISORY							
Planning and Organizing	2	2	0	0	0	0	2
Managing, Controlling/ and Delegating	2	3	0	0	0	0	2
Development of Others	2	2	0	0	0	0	2
TOTAL	20	20	20	20	20	20	20

Accurate performance appraisals are important. They provide feedback on an employee's contribution to the company, and the annual appraisal score is used to determine merit raises. When this appraisal system was introduced several years ago, evaluators attended extensive training sessions designed to ensure as much consistency and fairness as possible. Evaluators are also provided with detailed guidelines (Exhibit C). In spite of this, employee complaints continue to surface. Ann's job is to undercover any non-work-related biases in the evaluation process.

Concerns expressed by the employees at XYZ include the following.

- Evaluators are occasionally late with their evaluations. Rumors at the company suggest that the later the evaluation, the lower the score, as the manager may be attempting to postpone a controversial or hostile discussion with the employee about poor performance.
- There may be interdepartmental biases. Some managers may be attempting to raise their own department's status and salaries by artificially inflating the evaluation scores.
- Scores may be bumped up a few points to push a favored employee into the next higher raise category.
- Three-month probationary evaluations are considered unnecessary by the employees, because three months are generally not enough time to adjust to a new position or to reach the peak level of efficiency in a new job.
- There are rumors that the company is attempting to force early retirement by giving the long-time employees lower scores than others.

Ann has a sample of 144 employee evaluations conducted over the past calendar year; 10 variables were collected from each employee record as described below. The first 125 observations in the data set refer to employees who were still employed at XYZ at the end of the year; the remaining group left the company at some point during the year. The data available from these evaluations are contained in the file PERFEVAL. What should Ann's recommendations be?

EXHIBIT C   DETAILED EVALUATION GUIDELINES

EVALUATION GUIDELINES

This guide has been prepared to help you conduct the performance appraisal. The objectives are twofold:

1. General guidance (ideas to keep in mind when appraising performance).
2. "How-to's" of preparing the evaluation and conducting the session.

1. BE FAIR AND IMPARTIAL

Try to disregard bias about an employee. Concern yourself only with the facts and concrete instances of employee performance. Bias and prejudice may exist, but if you are aware of them, you will be in a good position to control them and their effects. A positive attempt must be made to be as fair as possible to the employee and to XYZ.

2. APPRAISAL PERFORMANCE

This is your role as a manager. Appraise performance against each accountability. It is not fair to base an evaluation, for example, on a person's educational attainments.

3. PERFORMANCE—NOT POTENTIAL

Accurate assessment of individual potential is important; however, the concern here is performance of the past year. Rate only the performance and do not rate what you think his/her potential is.

4. ACHIEVEMENT—NOT PROGRESS

Performance is rated in achieving results during a fixed time period. Success should be measured against previously agreed upon objectives.

5. APPRAISE PERFORMANCE FOR THE WHOLE YEAR

One month of commendable does not offset 11 months of satisfactory.

6. RATE EACH ACCOUNTABILITY INDEPENDENTLY

Do not let your performance rating on one area affect your rating against others. Try to rate each aspect of the evaluation independently.

7. CROSS RATINGS

Usually, evaluations are completed by the employee's manager. In some cases, another manager within XYZ may have important insights regarding the employee's performance; for example, if the employee worked under another manager for a good part of the year, it would be necessary to seek the views of the previous manager. Keep in mind, the goal is to have a complete, objective, and fair appraisal of performance.

8. BE A CONSCIENTIOUS RATER

Give a favorable rating only when the individual has merited it. Forget about giving people a "break." Often, this is difficult for us to accomplish. We want to be fair and often this means giving the benefit of the doubt. If we continue to do this over a period of time, we will find the standards we have set have strongly been lowered.

By lowering standards, XYZ's chances of attaining organization goals are weakened. Additionally, we hurt the individuals being evaluated by asking them to produce less to reach the same level. Be strict and fair to both parties. These two goals do not necessarily conflict.

9. MORE FREQUENT APPRAISALS

If an employee's performance needs improvement, establish a time frame for corrective action and re-evaluation. Discuss performance expectations, actions the employee can take to meet those expectations, and a specific time when you will re-evaluate the employee's performance.

CASE 44 PERFORMANCE EVALUATIONS

DEFINITIONS

SCORE	Evaluation score in points
LAG	Difference between the due date of the evaluation by the manager and the day on which the results were discussed with the employee, in days
EMP@EOY	Years with the company at the end of the year
POS@EOY	Years in the current position at the end of the year
TYPE	Type of evaluation (0 = 3 month evaluation, 1 = annual evaluation)
GRADE	Employee's grade scale: 1–7 are clerical or technical, 8–9 are skilled employees, 10–12 are supervisory, 13–17 are management
DEPT	Employee's department number
EDUC	Employee's education level: 12 is high school grad, 13 is one year training beyond high school (e.g., secretarial training), 14 is two years beyond high school (i.e., Associate's degree, nursing diploma), 16 is Bachelor's degree, 18 is Master's degree, 20 is PhD
EMP@EVAL	Years with company at time of evaluation
POS@EVAL	Years in the current position at the time of the evaluation

DATA SET

SCORE	LAG	EMP@EOY	POS@EOY	DEPT	GRADE	EDUC	EMP@EVAL	POS@EVAL	TYPE
184	−7	7.29	7.29	40	8	13	7.120	7.12	1
155	9	2.92	2.12	55	13	13	1.820	1.02	1
132	21	0.77	0.77	55	13	16	0.310	0.31	0
118	−7	0.80	0.80	55	4	13	0.230	0.23	0
102	15	13.78	4.70	41	2	12	13.160	4.07	1
178	4	9.65	9.65	64	3	12	9.020	9.01	1
115	2	2.19	2.19	41	9	16	2.150	2.15	1
128	13	0.50	0.50	62	8	16	0.290	0.29	0
99	35	12.76	2.73	62	2	12	12.100	2.07	1
133	−2	4.49	0.51	71	11	16	4.230	0.25	0
158	12	5.14	2.04	55	10	12	4.130	1.03	1
125	0	6.94	6.94	62	8	16	6.000	6.00	1
113	2	1.88	0.39	73	8	14	1.750	0.26	0
121	−12	6.35	0.28	73	17	18	6.290	0.22	0
172	10	3.78	3.78	72	8	16	3.030	3.03	1
159	49	11.70	11.70	64	8	14	11.670	11.67	1
135	14	6.10	3.17	55	4	13	5.830	2.90	1
155	7	7.59	5.29	55	10	18	7.320	5.02	1
128	22	0.56	0.56	41	2	13	0.310	0.31	0
127	62	3.42	3.42	73	10	16	3.170	3.17	1
134	45	0.31	0.31	62	8	16	0.370	0.37	0
126	13	0.77	0.44	55	2	12	0.290	0.05	0
131	16	4.24	4.24	41	10	18	4.050	4.05	1
136	−2	0.96	0.96	55	2	12	0.240	0.24	0
155	8	0.37	0.37	41	11	16	0.270	0.27	0
125	17	0.42	0.42	40	13	16	0.300	0.30	0

(continues on the following page)

SCORE	LAG	EMP@EOY	POS@EOY	DEPT	GRADE	EDUC	EMP@EVAL	POS@EVAL	TYPE
128	21	0.39	0.39	72	8	14	0.310	0.31	0
164	3	4.87	4.87	64	8	14	4.010	4.01	1
90	27	0.74	0.74	72	2	12	0.320	0.32	0
146	−5	4.28	2.79	73	11	14	3.470	1.99	1
159	19	0.92	0.28	55	14	20	0.950	0.30	0
125	12	4.80	4.80	74	8	16	4.040	4.04	1
129	36	3.42	3.42	62	8	14	3.100	3.10	1
149	4	2.61	1.28	63	8	16	2.340	1.01	1
182	−7	7.34	7.34	72	10	16	6.980	6.98	1
172	42	3.76	2.34	62	10	14	3.790	2.38	1
137	−11	6.52	0.28	71	17	16	6.470	0.22	0
171	30	3.16	2.75	62	3	12	2.500	2.08	1
144	34	1.27	0.73	55	3	12	1.090	0.55	1
163	28	1.39	1.39	61	8	14	1.090	1.08	1
165	4	6.89	5.12	74	8	16	6.750	4.98	1
128	11	6.52	5.39	72	3	12	6.170	5.03	1
133	6	1.29	1.29	63	8	14	1.020	1.02	1
162	−5	4.95	4.95	64	8	14	3.990	3.99	1
119	8	0.53	0.53	65	8	16	0.270	0.27	0
131	34	0.37	0.37	74	4	12	0.350	0.35	0
149	62	2.42	0.74	55	9	12	2.100	0.42	0
167	24	0.37	0.37	41	2	12	0.320	0.32	0
136	13	1.31	1.31	63	3	14	1.040	1.04	1
150	21	0.56	0.56	41	2	14	0.310	0.31	0
135	8	0.81	0.81	73	10	14	0.270	0.27	0
176	64	17.24	17.24	62	3	12	17.190	17.19	1
115	19	0.33	0.33	72	8	18	0.300	0.30	0
155	0	5.86	5.86	65	3	12	5.000	5.00	1
127	−12	1.17	1.17	62	3	13	0.970	0.97	1
61	37	4.72	0.28	62	6	14	4.790	0.35	0
115	15	3.16	0.86	55	2	13	2.580	0.28	0
121	23	1.69	0.71	55	2	14	1.290	0.31	0
160	−14	1.42	1.42	72	8	14	0.960	0.96	1
162	−1	12.58	3.99	72	13	14	11.590	3.00	1
155	4	2.86	2.86	72	8	14	2.010	2.01	1
124	−2	5.32	0.51	73	10	16	5.050	0.25	0
141	12	1.42	0.74	73	5	14	0.960	0.28	0
124	13	1.24	1.24	61	8	14	1.040	1.04	1
131	79	4.76	4.76	63	8	14	4.220	4.22	1
111	9	4.35	2.69	73	3	12	3.690	2.02	1
135	10	4.01	0.56	62	10	14	3.700	0.28	0
164	33	7.64	7.64	64	8	14	7.600	7.60	1
127	5	0.46	0.46	62	8	14	0.270	0.27	0
111	3	4.91	0.51	73	10	16	4.670	0.26	0
171	31	8.10	6.50	61	3	12	7.680	6.09	1
123	27	2.73	2.73	74	8	14	2.070	2.07	1

(continues on the following page)

CASE 44 PEFORMANCE EVALUATIONS

SCORE	LAG	EMP@EOY	POS@EOY	DEPT	GRADE	EDUC	EMP@EVAL	POS@EVAL	TYPE
128	15	0.94	0.94	62	8	16	0.290	0.29	0
128	−6	0.63	0.63	62	8	16	0.240	0.24	0
132	−12	0.88	0.88	70	6	16	0.210	0.21	1
173	34	8.73	8.73	63	8	14	8.100	8.10	1
126	36	0.90	0.54	72	3	12	0.720	0.35	0
120	8	2.96	1.13	62	8	14	2.020	0.20	0
121	30	0.96	0.96	65	3	12	0.330	0.33	0
111	12	0.92	0.92	74	8	14	0.280	0.28	0
129	58	0.96	0.96	61	3	14	0.410	0.41	0
132	9	2.52	2.52	71	10	16	2.070	2.07	1
75	10	10.90	6.50	55	10	16	10.920	6.53	1
157	6	12.76	12.76	73	4	12	12.020	12.02	1
155	20	7.39	3.12	73	13	16	7.330	3.05	1
152	−2	3.28	3.28	62	8	14	2.990	2.99	1
121	5	5.81	0.92	62	3	12	5.150	0.26	0
112	18	0.94	0.94	62	8	14	1.300	1.30	1
170	4	3.42	3.42	62	8	16	3.010	3.01	1
159	13	4.80	4.80	65	8	16	4.040	4.04	1
124	17	0.65	0.65	72	8	14	0.300	0.30	0
174	−7	3.44	3.44	62	8	16	2.980	2.98	1
120	9	2.44	1.13	74	13	16	1.580	0.28	0
139	0	1.96	0.89	55	10	14	1.310	0.24	0
139	0	2.76	2.76	72	2	12	2.000	2.00	1
160	−3	1.69	1.35	41	14	18	1.330	0.99	1
133	−3	5.35	5.35	72	8	14	4.990	4.99	1
147	14	7.56	7.56	64	8	16	7.560	7.56	1
126	15	0.52	0.52	61	8	18	0.290	0.29	0
112	−3	6.78	0.28	64	13	18	6.740	0.24	0
140	16	0.96	0.74	72	3	13	0.520	0.29	0
124	60	1.88	0.39	62	10	14	1.910	0.42	0
104	10	1.99	1.99	72	2	14	1.030	1.03	1
126	30	2.78	0.27	73	8	14	2.840	0.33	0
140	20	5.05	0.91	41	4	12	4.430	0.30	0
106	7	8.15	0.75	71	9	12	7.670	0.27	0
138	7	0.73	0.73	72	4	12	0.270	0.27	0
135	5	2.23	0.81	55	3	12	1.680	0.27	0
105	35	0.28	0.28	72	8	16	0.350	0.35	0
143	2	14.27	14.27	73	11	16	14.030	14.03	1
131	27	2.94	2.69	55	10	18	1.320	1.07	1
108	−10	11.28	0.28	74	17	14	11.220	0.22	0
171	−7	5.33	2.08	72	10	14	4.380	1.13	1
152	22	2.67	2.23	55	5	13	2.500	2.06	1
171	15	5.89	5.22	64	3	12	5.720	5.04	1
152	8	1.96	1.96	62	2	12	1.020	1.02	1
115	69	0.46	0.32	72	3	12	0.580	0.44	0
80	8	0.48	0.48	41	4	12	0.270	0.27	0

(concludes on the following page)

SCORE	LAG	EMP@EOY	POS@EOY	DEPT	GRADE	EDUC	EMP@EVAL	POS@EVAL	TYPE
115	15	2.13	0.36	41	5	16	2.060	0.29	0
134	−8	7.35	0.28	61	13	14	7.300	0.23	0
132	10	1.69	0.74	55	3	13	1.230	0.28	0
137	19	12.70	0.28	55	17	18	12.720	0.30	0
129	12	2.70	0.94	62	3	13	2.047	0.28	0
154	11	16.85	13.51	41	14	14	16.590	13.25	1
113	2	5.90	0.98	71	13	14	5.180	0.25	0
104	5	1.99	1.99	55	2	12	1.010	1.01	1
131	−4	1.10	1.10	72	8	14	0.240	0.24	0
117	34	0.81	0.81	62	8	14	0.350	0.35	0
171	−7	18.15	13.79	65	13	14	17.340	12.99	1
90	9	1.19	1.19	74	8	14	0.280	0.28	0
123	6	1.25	1.25	74	8	14	0.270	0.27	0
101	−3	8.25	4.49	62	13	16	7.760	3.99	1
40	87	0.96	0.96	62	8	18	0.480	0.48	0
117	6	2.80	2.80	74	8	14	2.020	2.02	1
132	9	1.27	1.27	74	8	14	1.020	1.02	1
159	−14	1.84	1.84	72	2	18	0.960	0.96	1
105	26	11.38	10.13	55	4	12	10.820	9.57	1
106	6	2.05	0.64	73	8	14	1.680	0.27	0
116	10	0.58	0.58	41	3	12	0.280	0.28	0
171	21	2.74	2.74	62	8	16	2.060	2.06	1
143	35	7.52	5.32	65	8	16	7.330	5.12	1
90	3	0.65	0.65	74	8	14	0.260	0.26	0
115	24	0.92	0.92	41	3	12	0.310	0.31	0
117	10	0.42	0.42	71	8	16	0.280	0.28	0

CASE 45

AMCHEM'S TAX CREDIT

As Harold Davies sat down to finish work on Amchem's tax return, he knew most of the details were available in the huge pile of files on his desk. There was one part, though, where he wasn't sure how to approach things—the sales tax on gas production.

Amchem manufactured chemicals in a plant in the Denver metropolitan area. A recent change in the Colorado tax law provided that all nonproduction use of power was subject to state sales tax. Harold knew that much of Amchem's use of natural gas *was* production-related. The problem was: how much? Public Service Company of Colorado bills and records revealed only the total gas consumption, not its eventual purpose. Harold knew that some part of that consumption went to heat the research facilities and other nonproduction and administrative buildings. That part of it, of course, depended on the weather, among other things, and would be subject to sales tax under the new law.

Harold suspected that most of the gas consumption went into the production processes for two groups of chemicals Amchem produced, labeled generically Group X and Group Y. The chemicals in Group X were fired in electric furnaces that use little gas (what little was used was used in other parts of the production process), while those in Group Y were fired in gas furnaces and used (Harold suspected) quite a lot more gas.

Harold felt that as long as he had a rational, defensible basis for his estimate of how much gas consumption was associated with the production process, he could get a credit of some kind for sales taxes paid on the total amount, and he felt it would be a substantial amount. Looking back over some of the files he'd requested, he found he

This case is based on a real business situation, though the names of the company and its products have been changed. The data were provided by David W. Howell.

had information and some quantities of presumed interest, for each month (MONTH) of the last five years (YEAR), as follows.

- Gas consumption (GAS): in hundred cubic feet (CCF), from gas company records.
- Heating degree days (DEGDAY): a measure of cold weather, obtained from the local weather bureau. The amount of gas required to heat buildings is expected to be related to this measure; that's why the weather bureau tracks it and provides it.
- X production: production of chemicals in group X, from Amchem's monthly records, in thousands of pounds.
- Y production: production of chemicals in group Y, from Amchem's monthly records, in thousands of pounds.

The data are available in file AMCHEM.

Harold knew that all these factors somehow contributed to overall gas consumption at Amchem. He figured other factors such as annual plant shutdowns in summer ("turn-around months") might affect things, too, but their effects were presumably represented by lower production during those months. One particularly extended shutdown in July of year 5 might have distorted things a bit, though.

As it got later in the afternoon, Harold wondered if there were some way to organize the information coherently. What he needed was an estimated amount of gas consumption for production use for year 5 that he could defend rationally. It would be even nicer if the method could be made routine, so it could be used in following years. Of course, it would also be possible to simply not claim the credit, but that was unappealing for any number of reasons.

DATA SET

YEAR	MONTH	DEGDAY	GAS	X	Y
1	1	1206	73950	13.6	23.5
1	2	942	70579	13.8	24.0
1	3	657	66408	15.4	23.2
1	4	421	50294	13.9	29.9
1	5	220	50582	13.9	26.9
1	6	55	45252	13.9	31.9
1	7	0	37145	13.8	14.2
1	8	6	43033	12.1	8.7
1	9	246	34696	9.2	8.6
1	10	460	53204	9.6	8.1
1	11	1017	51097	0.0	11.8
1	12	1070	48270	0.0	5.9
2	1	713	56711	4.5	9.7
2	2	780	60922	9.1	13.8
2	3	457	55902	10.7	13.2
2	4	359	44813	9.3	18.9
2	5	226	46704	7.8	19.4
2	6	12	32222	7.6	12.0
2	7	0	38513	6.3	12.7
2	8	2	39746	0.0	15.5

(concludes on the next page)

YEAR	MONTH	DEGDAY	GAS	X	Y
2	9	233	50892	6.1	19.2
2	10	561	48717	11.4	20.5
2	11	792	51416	10.7	16.9
2	12	1104	57815	7.6	18.5
3	1	1038	71074	13.9	21.7
3	2	770	69414	13.7	35.4
3	3	751	79210	13.8	34.6
3	4	338	65732	10.1	43.8
3	5	258	57291	10.7	36.1
3	6	36	53502	13.7	40.6
3	7	10	50987	14.1	46.2
3	8	39	36257	9.1	17.9
3	9	186	47561	13.6	25.9
3	10	493	60697	13.8	8.7
3	11	797	67631	13.1	21.8
3	12	1214	70093	14.0	23.4
4	1	1243	67630	15.0	21.9
4	2	892	68352	15.0	23.5
4	3	815	77922	15.0	31.8
4	4	453	62700	16.7	34.2
4	5	203	63142	17.8	33.8
4	6	15	53836	17.9	36.3
4	7	13	41154	13.4	26.4
4	8	6	54353	18.3	35.8
4	9	154	57456	18.1	37.0
4	10	412	72824	18.3	34.6
4	11	774	73905	19.8	37.9
4	12	963	70732	15.3	23.5
5	1	942	67563	7.8	9.4
5	2	1126	59740	14.0	24.9
5	3	657	52834	13.7	9.0
5	4	453	61719	13.3	8.8
5	5	218	57658	11.7	18.3
5	6	101	47576	13.6	18.9
5	7	0	12732	2.6	8.6
5	8	4	48714	6.6	12.7
5	9	186	51993	10.4	16.7
5	10	498	61752	10.7	10.2
5	11	710	58444	12.6	16.6
5	12	1233	62857	10.1	13.1

CASE **46**

NATURAL GAS PRICES

Your company produces natural gas and needs a statistical model to provide short-term forecasts of natural gas prices. Of course, forecasting the price of natural gas is difficult, because many different factors influence price. The demand for natural gas from the residential, commercial, and industrial sectors of the United States is a major factor. Natural gas prices may be seasonal, because, for instance, home heating requirements are higher in the winter months. Prices will also reflect the costs of producing natural gas. In addition, natural gas prices tend to mirror prices of *related* energy products such as coal and oil. Gas prices are also influenced by embargoes, production levels in other countries, domestic regulations, and other national and international political factors.

You have agreed to do an initial analysis of trends in natural gas prices. To accomplish this, you have been given monthly data on natural gas prices from January of 1984 to December of 1991. The numbers represent average wellhead price measured in dollars per thousand cubic feet. The data are in the file named GASPRICE.

The Energy Information Administration (EIA) maintains a website of the most current data on natural gas prices. This site is constantly updated as earlier published values are replaced with actual prices, and revisions are made to historical data. The latest data may be located by reviewing the EIA home page *www.eia.doe.gov/emeu/mer/contents.html*. The accuracy of your forecasts may depend on having the most current data available.

Construct a model or other statistical representation of natural gas prices and estimate prices for the first three months beyond the last month in your data. Prepare a report on your methodology and forecasts for presentation at a meeting next week.

Earlier data in this case were collected and analyzed by Thomas G. Funk. He found the data in the *Historical Monthly Energy Review 1973–1988* and the *Monthly Energy Review,* both by the Energy Information Administration, Department of Energy. Subsequent updates were located in the website indicated in the case.

DATA SET

Read across the row:

```
2.67  2.71  2.67  2.64  2.67  2.70  2.68  2.69  2.62  2.63  2.61
2.57  2.64  2.71  2.62  2.64  2.53  2.58  2.51  2.47  2.42  2.37
2.36  2.28  2.28  2.26  2.16  2.10  1.96  1.85  1.80  1.77  1.78
1.73  1.77  1.76  1.74  1.73  1.73  1.69  1.65  1.65  1.66  1.63
1.56  1.57  1.64  1.70  1.96  1.84  1.70  1.59  1.52  1.53  1.56
1.62  1.53  1.68  1.76  1.89  1.99  1.81  1.69  1.56  1.61  1.65
1.65  1.61  1.55  1.58  1.66  1.92  2.23  1.85  1.55  1.49  1.47
1.48  1.49  1.51  1.56  1.76  1.94  2.04  1.96  1.62  1.49  1.50
1.48  1.43  1.34  1.43  1.59  1.82  1.89  2.00  1.74  1.26  1.35
1.42  1.51  1.62  1.55  1.84  1.92  2.38  2.13  2.07  2.03  1.93
2.00  2.06  2.18  1.98  1.99  2.04  2.09  2.02  2.03  2.15  1.86
1.76  1.82  1.90  2.00  1.83  1.81  1.90  1.94  1.85  1.85  1.98
1.62  1.48  1.47  1.52  1.55  1.58  1.43  1.43  1.52  1.54  1.61
1.84  2.05  1.89  1.95  2.08  2.01  2.08  2.25  2.10  1.85  1.94
2.50  3.26  3.66  2.60  1.72  1.82  2.04  2.18  2.10
```

CASE 47

WESTAR BEVERAGE SALES

The Westar hotel, a 250-room business hotel located in a resort area of the Colorado Rockies, has two lounges on its premises: the Pour Nous Lounge, a pub-style lounge that seats about 50, and Gaieté, a 200-seat nightclub.

Jane Bartley is a staff analyst for Westar's corporate parent, charged (among other things) with predicting and analyzing beverage sales (liquor, beer, and wine, primarily) for the chain's business hotels. In fact, her deadline for producing some sort of forecast for Westar beverage sales was the end of the week. After years in the business, Jane had seen a lot of approaches to forecasting beverage sales, but none that struck her as particularly helpful. Most of them used only room occupancy as a basis for forecasting.

Not that it was unreasonable to consider room occupancy: clearly, the number of people staying in the hotel might well influence beverage sales. Further, hotel occupancy was (at least to some extent) predictable, as people made advance reservations, so it could be used for forecasting. In the case of the Westar, though, Jane suspected other things were involved, and the existing corporate forecasting methods (such as they were) made no allowance for them. Specifically Jane wondered if season, promotional food offered during happy hour, and the presence or absence of a piano player in the Pour Nous Lounge might be factors to consider. She was able to gather data for the most recent 30 months on some relevant variables (described on the next page). The data are available in file WESTAR.

The promotional food had been discontinued after month 12 and the piano player after month 13 as the previous management attempted to cut costs. Data from months 14 and 15 were unavailable because a change in ownership about then had led to some changes in record-keeping techniques (and the subsequent unavailability of certain records).

This case is based upon a real situation, described to us by Gail M. Derby, who provided the data. The names of the hotel chain and the lounges have been disguised.

Jane suspected that with these data she could develop a better understanding of the factors that really influenced beverage sales at the Westar and perhaps even influence her department to consider some new and better ways to do the forecasting.

DEFINITIONS

MONTH	Month number (month 1 was a January)
BEVTOTAL	Total beverage sales (in $), pub and nightclub
BEVPUB	Beverage sales (in $) for the Pour Nous Lounge
BEVNITE	Beverage sales (in $) for the Gaieté night club
FOODPUB	Amount spent on food (in $) for the Pour Nous Lounge
FOODNITE	Amount spent on food (in $) for the Gaieté night club
ROOMS	Total room occupancy for the Westar

DATA SET

MONTH	BEVTOTAL	BEVPUB	BEVNITE	FOODTOT	FOODPUB	FOODNITE	ROOMS
1	24443	7254	17189	2317	210	2107	4154
2	22924	7238	15686	2341	200	2141	4457
3	27501	8150	19351	3353	230	3123	5077
4	29282	8081	21201	2796	210	2586	4456
5	21447	7011	14436	2508	220	2288	4733
6	20404	6076	14328	2674	220	2454	5383
7	19051	8001	11050	2381	210	2171	5010
8	19235	8636	10599	2977	230	2747	5030
9	21498	9064	12434	1624	220	1404	5372
10	26619	8760	17859	1852	210	1642	4889
11	25600	7794	17806	1676	220	1456	4295
12	30373	8171	22202	2706	220	2486	3425
13	23483	8186	15297	2497	0	2497	3893
14	*	*	*	*	*	*	*
15	*	*	*	*	*	*	*
16	26004	7703	18301	2465	0	2465	4395
17	19624	6764	12860	2183	0	2183	4524
18	20459	6524	13935	3589	0	3589	5044
19	20956	4209	16747	3010	0	3010	5174
20	29754	7084	22670	3275	0	3275	6331
21	29098	5321	23777	3225	0	3225	4706
22	30559	5879	24680	3448	0	3448	4259
23	31038	5882	25156	3835	0	3835	4284
24	33833	4440	29393	3785	0	3785	3364
25	30095	6935	23160	3198	0	3198	4250
26	28163	4421	23742	2657	0	2657	4342
27	34051	5561	28490	3054	0	3054	5006
28	33139	6033	27106	2930	0	2930	4761
29	29468	5180	24288	1657	0	1657	3860
30	28896	6222	22674	2407	0	2407	5300

CASE 48

PRICE-FIXING CASES

Price-fixing (when two or more companies conspire to fix the price of a product or service rather than let it be determined by normal market forces) violates a number of federal laws in the United States, in particular, section 2 of the Sherman Act. Findings of guilty in such cases can lead to severe fines and other criminal penalties for the firms involved. One factor that influences the number of such cases in any given year is the actual number of price-fixing conspiracies, of course, but others include the zeal with which the Department of Justice prosecutes such cases and how it interprets both the law and the evidence. In a sense, then, we may think of the number of cases of price-fixing as the output of a complex process, a process by which society spawns such conspiracies and the government investigates and prosecutes them.

The data below are the number of price-fixing cases for each year from 1961 to 1987. This covers the administrations of Kennedy and Johnson (Democratic), Nixon and Ford (Republican), Carter (Democratic), and Reagan (Republican). The data include only those cases that resulted in either guilty findings or pleas of nolo contendere (which can be considered equivalent to guilty pleas for our purposes here). Each case is listed under the year in which an indictment was returned.

Do these data support the claim that the Reagan administration (1981–1988) adopted a more lenient policy with regard to prosecuting price-fixing cases than had been the case in previous administrations?

This case is based on data provided by E. W. Eckard, derived from the reports in the Commerce Clearing House Trade Regulation Reports.

DATA SET

YEAR	CASES	YEAR	CASES	YEAR	CASES	YEAR	CASES
1961	4	1968	8	1975	10	1982	3
1962	9	1969	1	1976	11	1983	9
1963	4	1970	6	1977	12	1984	9
1964	5	1971	8	1978	9	1985	2
1965	4	1972	12	1979	5	1986	5
1966	5	1973	6	1980	14	1987	7
1967	7	1974	10	1981	5		

CASE **49**

PROPERTY CRIMES

You've been asked to undertake a study of the determinants of property crimes in the United States for presentation at an upcoming meeting of business executives and community leaders. Your task is to provide evidence, for or against, common perceptions about property crime. Are crime rates higher in urban than rural areas? Does unemployment or education level contribute to property crime rates? How about public assistance? What other factors relate to property crimes? The file names PCRIMES may help answer some of these questions. Use these data to prepare a report on the characteristics of and determinants of property crimes in the United States.

DEFINITIONS

CRIMES	Property crime rate per hundred thousand inhabitants (property crimes include burglary, larceny, theft, and motor vehicle theft); calculated as # of property crimes committed divided by total population/100,000
PINCOME	Per capita income for each state
DROPOUT	High school drop out rate (%, 1987)
PRECIP	Average precipitation in inches in the major city in each state over 1951–80

The data for this case come from a variety of U.S. government sources, including: the 1988 Uniform Crime Reports, Federal Bureau of Investigation; the Office of Research and Statistics, Social Security Administration; the Commerce Department, Bureau of Economic Analysis; the National Center for Education Statistics, U.S. Department of Education; the Bureau of the Census, Department of Commerce and Geography Division; the Labor Department, Bureau of Labor Statistics; and the National Climatic Data Center, U.S. Department of Commerce. The data set was originally collected by Louis J. Moritz, an operations manager.

PUBAID   Percentage of public aid recipients (1987)
DENSITY   Population/total square miles
KIDS   Public aid for families with children, dollars per family
UNEMPLOY   Percentage of unemployed workers
URBAN   Percentage of the residents living in urban areas
STATE   1 = Alabama   26 = Montana
            2 = Alaska   27 = Nebraska
            3 = Arizona   28 = Nevada
            4 = Arkansas   29 = New Hampshire
            5 = California   30 = New Jersey
            6 = Colorado   31 = New Mexico
            7 = Connecticut   32 = New York
            8 = Delaware   33 = North Carolina
            9 = Florida   34 = North Dakota
            10 = Georgia   35 = Ohio
            11 = Hawaii   36 = Oklahoma
            12 = Idaho   37 = Oregon
            13 = Illinois   38 = Pennsylvania
            14 = Indiana   39 = Rhode Island
            15 = Iowa   40 = South Carolina
            16 = Kansas   41 = South Dakota
            17 = Kentucky   42 = Tennessee
            18 = Louisiana   43 = Texas
            19 = Maine   44 = Utah
            20 = Maryland   45 = Vermont
            21 = Massachusetts   46 = Virginia
            22 = Michigan   47 = Washington
            23 = Minnesota   48 = West Virginia
            24 = Mississippi   49 = Wisconsin
            25 = Missouri   50 = Wyoming

DATA SET

```
          CRIMES      DROPOUT     DENSITY      PRECIP           URBAN
STATE          PINCOME      PUBAID       KIDS        UNEMPLOY
  1     4003.1 12604    30.5    6.5    80.8   114  59.4    7.2   60.0
  2     4398.8 19514    26.4    4.3     0.9   593  53.2    9.3   64.3
  3     6861.2 14887    30.0    3.5    30.7   268   7.1    6.3   83.8
  4     3796.9 12172    21.4    5.9    46.0   190  49.2    7.7   51.5
  5     5705.7 18855    31.5    8.8   181.2   581  17.3    5.3   91.3
  6     5705.7 16417    24.0    3.7    31.9   317  15.3    6.4   80.6
  7     4642.2 22761    21.8    4.3   663.6   486  44.4    3.0   78.8
  8     4347.4 17699    29.0    4.3   341.7   267  41.4    3.2   70.6
  9     7819.9 16546    36.5    4.1   227.8   240  55.2    5.0   84.6
 10     5661.2 14980    35.0    6.4   109.2   252  48.6    5.8   62.4
 11     5731.9 16898    15.5    4.9   170.9   481  23.5    3.2   86.5
 12     3738.2 12657    20.4    2.7    12.2   250  11.7    5.8   54.0
 13     4810.4 17611    22.2    7.5   208.7   309  33.3    6.8   83.3
 14     3770.0 14721    24.1    3.6   154.6   263  39.1    5.3   64.2
```

(concludes on the next page)

	CRIMES	PINCOME	DROPOUT	PUBAID	DENSITY	KIDS	PRECIP	UNEMPLOY	URBAN
STATE									
15	3819.8	14764	13.4	5.0	50.6	348	38.6	4.5	58.6
16	4514.7	15905	15.9	3.8	30.5	338	28.6	4.8	66.7
17	2804.7	12795	32.1	7.0	93.9	204	43.6	7.9	50.9
18	5043.3	12193	38.4	8.8	99.0	167	59.7	10.9	68.7
19	3420.3	14976	21.2	6.5	38.9	370	43.5	3.8	47.5
20	4897.9	19316	23.5	5.1	469.9	329	41.8	4.5	80.3
21	4371.3	20701	24.0	5.9	752.7	536	43.8	3.3	83.8
22	5342.7	16387	28.6	8.5	162.2	481	31.0	7.6	70.7
23	4024.7	16787	11.3	4.6	54.1	515	26.4	4.0	66.9
24	3267.6	10992	34.4	11.1	55.5	119	52.8	8.4	47.3
25	4292.2	15492	23.9	5.5	74.6	264	33.9	5.7	68.1
26	4144.0	12670	15.5	4.5	5.5	362	11.4	6.8	52.9
27	3866.8	15184	13.3	3.8	20.9	320	30.3	3.6	62.9
28	5672.5	17440	18.7	2.6	9.6	273	4.2	5.2	85.3
29	3186.1	19016	25.4	1.7	120.7	407	36.5	2.4	52.2
30	4712.5	21882	20.3	5.6	1033.9	357	41.9	3.8	89.0
31	5948.1	12481	26.8	5.7	12.4	225	8.9	7.8	72.1
32	5212.0	19299	33.3	8.0	378.0	523	39.3	4.2	84.6
33	4360.4	14128	30.9	5.0	132.9	243	42.5	3.6	42.9
34	2668.9	12720	11.6	3.1	9.6	350	15.4	4.8	48.8
35	4193.2	15485	20.8	7.5	264.7	298	37.8	6.0	73.3
36	5154.6	13269	24.2	4.8	47.2	278	30.9	6.7	67.3
37	6513.0	14982	29.2	4.0	28.8	348	37.4	5.8	67.9
38	2814.4	12168	18.9	6.1	267.4	347	40.0	5.1	69.3
39	4807.7	16793	28.0	6.0	940.9	450	41.9	3.1	87.0
40	4671.2	12764	32.2	6.3	114.9	186	51.6	4.5	54.1
41	2467.3	12475	13.1	3.9	9.4	270	17.5	3.9	46.4
42	3936.7	13659	32.8	6.4	118.9	155	48.5	5.8	60.4
43	7365.1	14640	34.1	4.5	64.3	169	42.0	7.3	79.6
44	5335.5	12013	17.5	3.2	20.6	343	15.3	4.9	84.4
45	4098.2	15382	17.3	5.6	60.1	469	33.7	2.8	33.8
46	3877.5	17640	23.4	4.0	151.5	257	45.2	3.9	66.0
47	6646.6	16569	21.9	5.8	69.9	443	38.6	6.2	73.5
48	2107.4	11658	22.9	8.2	77.8	238	40.7	9.9	36.2
49	3757.6	15444	16.3	7.7	89.2	473	30.9	4.3	64.2
50	3653.1	13718	20.4	3.2	4.9	303	13.3	6.3	62.7

CASE **50**

ROCKVILLE PHYSICIANS' GROUP

In late November, Dr. Eaton sat down to review the results of a study of cesarean sections (C-sections) at the Rockville Physicians' Group (RPG). He knew there were interesting findings in the data gathered over the past year, and he wanted to understand those findings. For example, he had discovered that the C-section rates differed dramatically among the various RPG physicians, from about 14% of all births to about 30%. He also knew, though, that how he summarized and communicated those findings to the obstetrics/gynecology (OBG) staff at a meeting scheduled for mid-January was probably just as important as the results themselves, at least in terms of getting them to "buy in" to the ideas of quality monitoring and measurement that he wanted to establish at RPG.

RPG was founded in the early 1950s as one of the earliest group practices in its geographic region. Located 30 miles from a major city, Rockville was at that time perhaps best known as an attractive suburb and as the home of a state university. Since then, it had grown enormously to about 100,000 residents, plus more than 10,000 students. Other industrial and governmental offices were located in or near Rockville and a major hi-tech firm had its laboratories in a university-sponsored research park near town. Thanks to its attractive scenery, generally liberal attitudes, and a far-seeing (aggressive?) no-growth policy, Rockville had become an affluent (exclusive?) city, and RPG had prospered along with it. It was easily the largest such practice in the Rockville area and had contracts and arrangements with most major insurance and reimbursement agencies, various health maintenance organizations, and others. In all, 50 physicians worked at the center. Total other staff ran about 185.

Dr. Eaton had joined RPG more than 30 years earlier, after leaving the armed services, and practiced internal medicine there until recently, when he retired from active practice

This care is based on a real situation, though the names of the organization and the individuals have been changed. The situation was described for us by John S. Avery, M. D.

and took an (approximately) half-time job as RPG's medical director. As one of his first major forays into administration, he had played a major role, including negotiations with contractors and various regulatory agencies, in the design and construction of a building expansion for the group. He had been involved in contract talks with some of the insurance and other reimbursement agencies, and his interest in computers and data management had grown in the last few years. He had discovered he enjoyed administration, and felt he'd been reasonably successful at it.

There had been a quality assurance committee for some years at RPG, chaired by one of the physicians. Historically, topics for study had been selected in a more-or-less democratic manner by the physicians involved, but those topics had tended to be administrative rather than medical (studies of waiting times, telephone systems, and so on). Dr. Eaton felt the Group needed to be systematically involved in quality monitoring of the medical aspects of their operations, beyond the usual kinds of case review sessions they routinely held. In his words, he was

> hoping to build into our quality assurance efforts . . . these [attitudes and philosophies]:
>
> • Focus on the center of a quality/frequency bell curve, moving the curve to the right, rather than on the detection of instances of poor care, lopping off the left tail of the curve.
> • Encouraging self-examination, becoming aware of what our patterns of practice actually were. Stepping back and looking at ourselves. Escaping habits of which we were scarcely conscious.
> • Encouraging a research attitude, being quantitative and systematic in looking at ourselves. Creating the possibility of publication.
> • Having the courage to be independent; learning from ourselves, not being bound by the received opinions of the day.
>
> I saw this as a way to enhance the fun of practice as well as the quality of care, to lessen the insidious onset of staleness and routine that you see in all professions.

Accordingly, the ground rules of the committee were changed to require that the topics be restricted to the study of the patterns of care and the use of the results in medical decision making. This year, after some discussion with various people involved, the committee decided to focus on the OBG department, and particularly the study of C-sections. Dr. Eaton set up a small team working under his supervision to gather and process data. The report from this study was to be presented at the mid-January meeting.

Cesarean section rates in the United States have been running about 20% to 25% of births in recent years. This rate is a concern to many in the medical community, both because of the costs involved and because it is dramatically higher (double, in many cases) than the rate in the rest of the developed world. There is perhaps some evidence that it is influenced by fear of clinical failure and/or malpractice claims, but Dr. Eaton didn't know specifically whether that or any other identifiable influence was at work in the case of the medical center, whose rates had been around the national averages in recent years.

Working with Mary Spiegel, an assistant, he identified all births, including C-sections and vaginal deliveries, from October 1 to September 30. From the computer system used for administrative and billing operations, they were able to identify the following information for all the cases.

- Chart number.
- Procedure code.
- Doctor number (six different physicians were involved).
- Date of service.

- Charge.
- Financial code (type of insurer, or none).
- Diagnosis code.

In addition, they identified a sample consisting of all C-sections (178 of them) and 176 randomly chosen vaginal deliveries for more detailed analysis. For the sample of 354 patients, charts were pulled and the following further information was recorded.

- Maternal age.
- Parity.
- Gravida status.
- Immediate and underlying reasons for the C-section.
- Potential reasons for a C-section (this refers to cases in which a C-section was not performed, but where some mention was made of a condition potentially leading to a C-section).
- Time in labor.
- Apgar-1 and Apgar-5 Scores (a measure of general condition at age one minute and five minutes for the baby).
- Gestational age.
- Birth weight.
- Maternal death (Y/N).
- Fetal death (Y/N).

While it was a time-consuming exercise, there had been no particular difficulties in obtaining the information, at least none that Mary (with occasional consultation with Dr. Eaton) had been unable to resolve. The process had been greatly facilitated by the existence of fairly standard (nationwide) forms and coding schemes for OBG procedures.

The combined data had been entered into a database package for a personal computer. A sample page of direct printout from the program is attached (Exhibit A). Using the database package, it was possible to select certain subsets of the data based on user-specified criteria. For example, it was possible to ask to look at all those (and only those) cases in which labor time was greater than two hours and the Apgar-1 score was less than 5, if such a combination was considered interesting. It was also possible to extract data from the data base package and "send" it to spreadsheet programs or graphics packages.

The Quality Assurance and Utilization Management (QAUM) committee consisted of the RPG's administrator, nursing supervisor, the head nurse of the area involved (OBG, in this case), and the area physicians. All were expected to have reviewed the agenda and materials for the meeting beforehand and everyone was expected to comment on the materials or the topic generally. At the meeting scheduled for mid-January, the topic would be the results of the C-section study. Dr. Eaton's original goals for the study had been basically to provide information to the staff, help them understand what the data suggested, elicit their suggestions for changes that might be made, and obtain their support for the monitoring and quality activities he hoped to continue at RPG. Now armed with the results of the C-section study, he wondered how best to present them to the OBG staff.

EXHIBIT A   SAMPLE PAGE OF PRINTOUT

```
    100% indexed          44632 Records indexed
. display structure
Structure for database: C:
Number of data records:    1017
Date of last update   : 03/14/91
Field  Field Name   Type        Width    Dec
    1  CHART        Character     6
    2  PROC_CODE    Character     5
    3  DOCTOR_NUM   Character     4
    4  AGE          Numeric       2
    5  PARITY       Numeric       2
    6  GRAVIDA      Numeric       2
    7  DIAG_CODE    Character     5
  · 8  CLASS        Character     1
    9  BIRTH_WGT    Numeric       4        1
   10  GEST_AGE     Numeric       2
   11  REASON_1     Numeric       2
   12  REASON_2     Numeric       2
** Total **                      38

Command Line    :<C:>:                    :Rec: 1/1017       :Ins    :Num
                     Enter a dBASE III PLUS command.
```

Handwritten annotation: Obtained by client-pull, not from main computer.

Record#	CHART	PROC_CODE	DOCTOR_NUM	AGE	PARITY	GRAVIDA	DIAG_CODE	CLASS	BIRTH_WT	GEST_AGE	REASON_1	REASON_2
1	313560	59410	103	33	1	3	V222	2	6	40	0	0
2	351602	59500	103	39	2	4	V222	1	9	39	11	18
3	222600	59410	18	34	1	2	V222	9	8	50	0	0
4	307185	59500	42	30	1	3	6612	1	E	40	22	0
~~5~~	~~307185~~	~~59500~~	94	0	0	0	6612	1	C	0	0	0
~~6~~	~~349227~~	~~59410~~	42	0	0	0	V222	1	C	0	0	0
7	349135	59500	42	27	0	1	V222	1	0	40	24	0
~~8~~	~~349135~~	~~59500~~	80	0	0	0	7684	1	C	0	0	0
~~9~~	~~350727~~	~~59410~~	103	0	0	0	V222	8	C	0	0	0
10	349645	59410	80	21	0	1	V222	1	7	42	0	0
11	335661	59410	18	31	1	2	V222	1	E	40	0	0
~~12~~	~~340738~~	~~59410~~	18	0	0	0	V222	7	C	0	0	0
~~13~~	~~342832~~	~~59410~~	80	0	0	0	V222	1	C	0	0	0
~~14~~	~~255875~~	~~59410~~	103	0	0	0	V222	1	C	0	0	0
15	342472	59410	18	28	2	4	V222	9	7	37	0	0
16	336306	59410	80	29	1	2	V222	1	E	40	0	0
17	293233	59500	103	36	0	3	V222	2	7	39	14	23